The anatomy of the d

B. F. Kaupp

Alpha Editions

This edition published in 2024

ISBN : 9789366388588

Design and Setting By
Alpha Editions
www.alphaedis.com
Email - info@alphaedis.com

As per information held with us this book is in Public Domain.
This book is a reproduction of an important historical work. Alpha Editions uses the best technology to reproduce historical work in the same manner it was first published to preserve its original nature. Any marks or number seen are left intentionally to preserve its true form.

FOREWORD

Advanced work in the study of poultry husbandry is now being done in this country, to a greater or less extent, at all the two score and more Agricultural and other Colleges and Experiment Stations. From these institutions comes the demand for a text-book on the Anatomy of the Domestic Fowl. No complete text-book on the subject, up to the present, has existed. It is with the hope of meeting the demand that this book is published.

In supplementing the information gathered from the books and articles listed in the bibliography, the author has spent much time and effort in obtaining the matter here presented. As this, however, is the initial complete text on the subject necessarily much remains to be added and corrected. The author would welcome suggestions and corrections from any one into whose hands the book may come.

B. F. KAUPP.

THE NORTH CAROLINA STATE COLLEGE OF AGRICULTURE,

WEST RALEIGH, NORTH CAROLINA,

September, 1918.

OSTEOLOGY

Osseous Structure.—Bone is structurally modified connective tissue which has become hard by being impregnated with calcium salts.

Kinds of Bone Tissue.—There are two kinds of bone tissue: substantia compacta, or compact bone tissue; and substantia spongiosa, or spongy, cancellous bone tissue.

Compact Bone Tissue.—The compact bone tissue forms the hard outer layer of all bones. It is thickest in the shaft and becomes thin toward the extremities. Through the compact bone tissue approximately parallel with the longitudinal axis of the bone, run canals called *Haversian canals*, through which pass blood and lymph vessels for the nourishment of the bone and nerves. The Haversian canals are surrounded by *concentric lamellæ*. The spaces between the cylinders thus formed are filled with *interstitial lamellæ*; and both the exterior surface of the bone and the interior surface surrounding the medullary canal, are built up of peripheral, or circumferential lamellæ. Between the lamellæ, somewhat irregularly placed, are minute reservoirs, called *lacunæ*, which contain bone corpuscles. From the lacunæ radiate minute canals, or *canaliculi*, which maintain circulation through the bone substance, and which communicate with the Haversian canals. Complex anastomoses exist among the canaliculi. Still other channels for the passage of blood-vessels are Volkmann's canals which pierce the peripheral lamellæ, thus allowing vessels to pass from the periosteum to the Haversian canals. Similar channels afford communication between the inner Haversian canals and the medullary cavity.

The entire structure composed of an Haversian canal, its surrounding lamellæ, lacunæ, and canaliculi, with their contained vessels, is called an *Haversian system*.

Cancellous Bone Tissue.—The cancellous bone tissue forms the bulk of the short, flat, and irregular bones and of the extremities of the long bones. It consists of delicate bony plates and spicules, which intercross in various directions. The spaces between these plates and spicules, called *cancelli*, are occupied by marrow except in the bones that are pneumatic. The blood-vessels, lymphatics, and nerves course through this marrow but are not arranged in an Haversian system.

FIG. 1.—Longitudinal section of compact bone of the femur of the hen. 1, Haversian canals. 2, Lacunæ with their canaliculi.

The Periosteum.—Covering the surface of bone, except at the articular surface where it is covered with cartilage, is a membrane, the periosteum, which consists of two layers: an outer, fibrous, protective layer, and an inner, cellular, osteogenic layer. The outer layer consists principally of white fibrous tissue. The inner layer contains many more connective-tissue cells, which gradually become more closely aggregated as we proceed toward the osseous surface; but there is no sharply defined line of demarcation between the two periosteal layers.

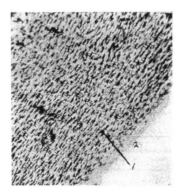

FIG. 2.—Transverse section of compact bone of the femur of the hen. 1, The lacunæ and canaliculi. 2, The periosteum.

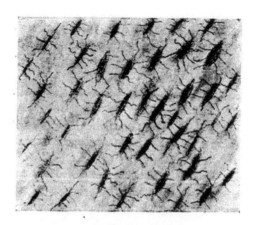

FIG. 3.—Transverse section of compact bone of the femur of the hen showing the lacunæ and canaliculi under high magnification.

The periosteum is firmly attached to the bone by trabeculæ of fibrous tissue, called the *fibers of Sharpey*. These fibers of Sharpey penetrate the bone at right angles to its surface and carry blood-vessels.

Marrow.—There are two kinds of marrow: yellow, or medulla ossium flava, and red, or medulla ossium rubra.

The *yellow marrow* occurs in all bones except the femur and proximal portion of the tibia of adult fowls. It is composed of a network of fibrous tissue carrying blood-vessels, fat cells, and myelocytes, or marrow cells.

The *red marrow* is found throughout the femur and the proximal portion of the tibia, and in a few of the pelvic bones and vertebræ in the adult fowl, and in certain other bones of the baby chick. Red marrow consists of a delicate network of connective tissue supporting a dense capillary plexus, a small amount of fat, and numerous cells. The cellular elements of red marrow consist of marrow cells which contain large nuclei and possess ameboid movement, red blood cells, giant cells containing one or more nuclei, and various kinds of leucocytes, including eosinophiles, mast cells, and also osteoclasts.

Growth of Bone.—In the baby chick, only the shaft and a portion of the extremities of the long bones are thoroughly ossified, the extreme ends, and of the femur most of the articular head, being cartilaginous. The bones grow in length by an increase in the cartilage, the cartilage gradually becoming ossified. Growth in diameter is accomplished by the constant deposition of new layers of bone beneath the periosteum. During this process the osteoclasts absorb the bone from within. The formation of the marrow cavity is thus effected.

Classification of Bones.—The bones of the fowl are classified as long, short, flat, and irregular.

Long Bones.—The long bones occur in the legs and wings, where they serve as levers to sustain weight and make locomotion possible. A long bone consists of a shaft and two extremities. The superior is called the *proximal* and the inferior the *distal extremity*. The expanded articular surfaces in forming joints with adjoining bones afford ample space for the attachment of ligaments. The shaft is cylindrical and hollow.

Short Bones.—Short bones occur in the feet and in the wings. Their structure is similar to that of the long bones.

Flat Bones.—The flat bones occur where extensive protection is needed, as in the cranial region; or where large surface for muscular attachments is needed, as in the costal and pelvic regions. Flat bones are made up of two thin layers of compact bone with a variable amount of cancellous tissue interposed.

Irregular Bones.—The irregular bones include the vertebræ, the patellæ, and the carpal bones.

Composition of Bone.—Bone consists of organic and inorganic matter. Organic matter gives toughness and elasticity to the bone, and inorganic matter hardness. The organic substance of bone is called ossein. When boiled in water ossein is resolved into gelatin. The following tables[1] give the results of an analysis of the femur, fresh, of a mature hen.

1. Grateful acknowledgment is hereby made to Dan M. McCarty, Chemist, Animal Industry Division, North Carolina Agricultural Experiment Station, for this analysis.

Fresh femur:	
Water	18.23 per cent.
Dry matter	81.77 per cent.
Dry matter:	
Organic matter	63.09 parts
Inorganic matter	18.68 parts
Salts in dry matter:	
Calcium	6.970 per cent.

Magnesium	0.283 per cent.
Potassium	0.004 per cent.
Sodium	0.276 per cent.
Iron	0.020 per cent.
Phosphorus	3.210 per cent.
Sulphur	0.085 per cent.
Chlorine	0.520 per cent.
Carbon dioxid	0.550 per cent.

The inorganic matter of the femur of the hen consists of 18.68 parts or 22.84 per cent. of dry matter, and the organic matter of 77.16 per cent. Stated in other words the femur, including its contained marrow, consists of organic and inorganic matter in the ratio of 3.4, approximately, to 1.

The Skeleton of the Fowl.—The skeleton of a bird is remarkable for the rapidity of its ossification. It is worthy of note that other parts of the bodies of adult birds also become ossified. Among such parts are the tendons of the muscles of the legs, of the feet, and of the neck; the plates of the corneal margin of the sclerotic tunic of the eye; and the stapes of the ear. Ossification in birds at the attachments of the semi-lunar valves of the aorta and of the pulmonary artery has been reported by Owen.

The bony structure is compact, and the bones contain a greater proportion of phosphate of lime than do the osseous structures of mammals. Especially is this the case in those parts of the skeleton which are permeated by air.

THE DIVISIONS OF THE SKELETON

(The Skull (Cranium

  ((Face

 The Axial skeleton ((Cervical region

  ((Dorsal region (Ribs

(The Vertebral column ((Sternum

(Lumbar region

(Sacral region

(Coccygeal region

(Scapula

(Shoulder girdle (Coracoid

((Clavicle

(

((Arm (Humerus
((
(Fore limb (Forearm (Radius

(((Ulna

((

(((Carpus

The Appendicular skeleton ((Hand (Metacarpus

((Phalanges
(
(Pelvic girdle (Ilium
((Hip bone)   (Ischium
((Pubis
(

((Thigh (Femur

(Hind limb (

(Leg (Tibia
((Fibula

(
(Foot (Metatarsus

(Phalanges

The bodies of birds contain many air reservoirs to make them light that flying may be more easy. Many bones have their weight in proportion to size and strength thus greatly reduced. In very young birds the cavities of bones contain, instead of air spaces, loosely arranged red marrow, which is in most bones later absorbed. The air reservoirs in bones are most capacious in the best flyers. In the non-flyers more of the bones retain their red marrow.

The bones supplied with air spaces are relatively larger than in mammals, and are provided with small transverse osseous columns which cross in different directions and from side to side. These cross beams give stability to the thin wall of the bone. The membranes lining these cavities are very vascular.

The Axial Skeleton
THE SKULL

The skull is divided into the cranial and facial portions. In these parts we find present 31 bones: one occipital, two parietal, two frontal, one ethmoid, one sphenoid, and two temporal; all of which constitute the cranial group; two premaxillary, two maxillæ, two nasal, two lacrimal, two palatine, two pterygoid, two zygomatic, one vomeral, the two jugal, and two quadrato-jugal, which constitute the facial group; two quadrati and one inferior maxillary, which constitute the inferior jaw group.

The peculiarities of the skull are the long os incisivum and the single condyle located on the occipital bone just below the foramen magnum. The condyle articulates with the atlas.

The head of the bird is small in proportion to the size of the body, and in front it is conical in shape.

THE CRANIUM

The cranial cavity, or cavum cranii, incloses the brain with its membranes and vessels.

The dorsal wall, or roof, is formed by the frontal and the parietal bones. In the median line of the cerebral portion is the internal parietal crest. The roof of the cerebellar portion is marked centrally by a groove.

The posterior wall of the cerebellum is formed by the occipital bone.

The lateral wall is formed chiefly by the temporal bone. It is marked by a ridge which divides the cavity into the cerebral and cerebellar compartments. The cerebral portion is marked by a depression which receives the optic lobes. A crest divides this cavity from the optic portion. The walls are marked by digitations and vascular grooves.

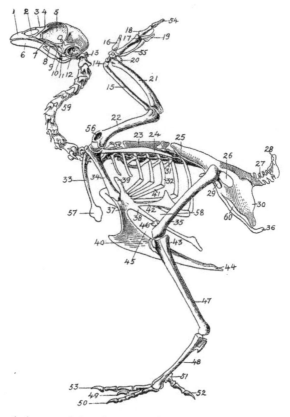

FIG. 4.—The skeleton of the domestic fowl. 1, Os incisivum. 2, External nasal opening. 3, Os nasale. 4, Os lachrymale. 5, Lamina perpendicularis. 6, Os dentale. 7, Os palatine. 8, Os quadrato-jugal. 9. Os pterygoideum. 10, Os quadratum. 11, Os articulare. 12, External auditory canal. 13, Atlas. 14, Os carpi radiale. 15, Radius. 16, First finger. 17, Os metacarpus. 18, Second finger. 19, Third finger. 20, Os carpi ulnare. 21, Ulna. 22, Humerus. 23, Thoracic vertebræ. 24, Scapula. 25, Os ilium. 26, Foramen ischiadicum. 27, Caudal vertebræ. 28, Pygostyle. 29, Foramen obturatum. 30, Os ischium. 31, Processus uncinatus. 32, Vertebral rib. 33, Os claviculum or furculum. 34, Os coracoideum. 35, Os femoris. 36, Os pubis. 37, Body of sternum. 38, Lateral internal process of sternum. 39, Costal process of sternum. 40, Sternal crest, cristi sterni, or keel of breast-bone. 41, Sternal rib. 42, Lateral external process of sternum. 43, Os fibula. 44, Xyphoid process of sternum. 45, Internal notch. 46, Os patella. 47, Os tibia. 48, Os metatarsus. 49, Second toe. 50, Fourth toe. 51, Os metatarsale. 52, First toe. 53, Second toe. 54, Second phalanx of second finger. 55, Os metacarpus. 56, Air opening in humerus. 57, Hypocledium. 58, External notch. 59, Cervical vertebræ. 60, Foramen oblongum.

The ventral wall, floor, or basis cranii interna, may be regarded as forming five fossæ. These are: one anterior, one middle, one posterior, and two lateral. The *anterior fossa* supports the frontal and olfactory parts of the cerebrum. It lies at a higher level than the middle fossa. The *middle fossa*, or *fossa cranii media*, is circular in outline and extends into the lateral fossæ which lodge the olfactory lobes. Just back of the middle fossa is the sulla turcica, upon which rests the pituitary body. The *posterior fossa*, or *fossa cranii posterior*, lodges the medulla oblongata.

Bones of the Cranium (Figs. 4, 6, 7, and 8).—The bones of the cranium fuse early in the chick's life. The sutures uniting the bones can usually be seen in the fetus or in the baby chick soon after it emerges from the shell. The major portion of the cranial bones become entirely fused. Each cranial and each facial bone ossifies from a distinct center or centers.

The cavity for the cerebrum is much larger than the cavity for the cerebellum. The cranial cavities in birds are relatively larger than in mammals. The bones are designated as in mammals. There are three single bones, the occipital, the ethmoid, and the sphenoid. Those in pairs are the frontal, the parietal, and the temporal.

The Occipital. *Location.*—The occipital bone or os occipitale, is situated at the posterior part of the cranium, of which it forms the posterior wall. This part is called the base of the cranium. The occipital bone articulates with or touches inferiorly, the sphenoid, laterally, the temporal, and superiorly, the parietal.

Development.—The occipital bone is developed from four centers of ossification; the dorsal, or os occipitale superius, two lateral, or ossa occipitales laterales, and the ventral, or os occipitale inferius, all of which may be seen distinct in the baby chick (Fig. 5, Part II, No. 1).

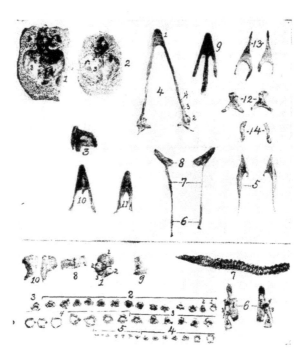

FIG. 5.—Bones of the head and vertebra.

Part I.—1, The cranium. 2, The skull cap. 3, The ethmoid bone. 4, The inferior maxilla. 5, The palatine bone. 6, The quadrato-jugal. 7, The jugal. 8, The superior maxilla. 9, The premaxilla. 10, The horny covering for the premaxilla. 11, The horny covering for the os dentale. 12, The os quadratum. 13, The nasal bone. 14, The pterygoid.

Part II.—1, The os occipitale (1, dorsal; 2, two lateral; 3, ventral portions and 4, foramen magnum). 2, The fourteen cervical vertebræ. (1, atlas; 2, axis). 3, The seven dorsal vertebræ. 4, The fourteen lumbo-sacral vertebræ. 5, The seven coccygeal vertebræ. 6, The pelvis (1, ilium; 2, ischium; 3, pubis; 4, cotyloid cavity or acetabulum). 7, The vertebral column complete from the baby chick. 8, The parietal bone. 9, The temporal bone. 10, The frontal bone.

Description.—In the center of the occipital bone is the *foramen magnum*. Through this foramen the spinal cord extends into the cranial cavity and connects with the medulla oblongata. The occipital bone has a single condyle, which is located just below the foramen magnum and articulates with the atlas. At the base of the condyle a small *subcondyloid fossa* receives the body of the atlas during extreme flexion of the head. In the center of the lateral wing of the occipital bone there is a small foramen through which passes the hypoglossal nerve. Somewhat laterally from this foramen there is an opening through which the vagus, or pneumogastric nerve passes. Laterally from these is the *canalis caroticus et jugularis*. Between the os occipitale superius, or dorsal portion, and the ossa parietalia is a space to which ligaments are attached, called the fontanel.

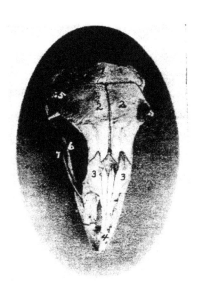

FIG. 6.—View of the frontal region of the skull of a hen. 1, Parietal. 2, Frontal. 3, Nasal. 4, Premaxilla. 5, Temporal. 6, Inferior jaw. 7, Jugal bone.

The Ethmoid. *Location.*—The ethmoid, or os ethmoidale, forms the anterior wall of the cranium, and the orbital septum. It is related anteriorly with the vomeral, superiorly with the nasal and frontal, posteriorly with the temporal, and inferiorly with the sphenoid, and palatine.

Description.—The ethmoid consists of a perpendicular and a horizontal lamina. The perpendicular lamina, located between the orbital cavities, is called the *septum interorbitale*. On each side of the septum interorbitale and near the superior orbital roof are two foramina for the passage of the olfactory nerves. In the horizontal plate, which forms the anterior cranial wall, are located the *optic foramina* through which pass the optic nerves.

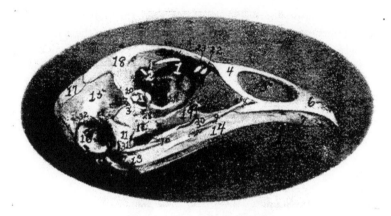

FIG. 7.—Side view of the skull of a hen. 1, Lamina perpendicularis. 2, Foramen for the passage of the nerve of smell. 3, Foramen for the passage of the optic nerve. 4, Nasal bone. 5, External nasal opening. 6, Premaxilla. 7, Os dentale. 8, Superior maxilla. 9, Os jugal. 10, Os quadrato-jugal. 11, Os quadratus. 12, Pterygoideus. 13, Os articulare. 14, Inferior maxilla. 15, Temporal. 16, External auditory canal. 17, Parietal. 18, Frontal. 19, Palatine. 20, Orbital process, posterior to which is the processus zygomaticus. 21, Interorbital foramen and optic foramen for passage of optic nerve. 22, Attachment for inferior oblique; 23, for superior oblique; 24, for levator palpebræ superioris; 25, for internal rectus; 26, for superior rectus; 27, for external rectus; 28, for inferior rectus; 28, for inferior rectus; 29, for depressor palpebræ inferioris; 30, for orbicularis palpebrarum; 31, for tensor tympani; 32, for circumconcha.

The Sphenoid. *Location.*—The sphenoid, or os sphenoidale, forms the floor of the cranial cavity and articulates with or touches posteriorly the occipital, laterally the temporal, and anteriorly the palatine, and ethmoid.

Description.—The sphenoid bone, the largest part of the cranial floor, is formed by the fusion of the nasal and the cranial parts. It is a three-cornered bone with two, thin, broad wings. These wings are divided into two portions, orbital wings, or *alæ orbitales*, and temporal wings, or *alæ temporales*. The temporal wing forms a cover for the Eustachian tube trumpet, and for the canal coming from the sella turcica, which latter gives passage to the intracranial carotid artery. The orbital wing forms the lower portion of the posterior wall of the orbital cavity, and lies directly before the os petrosum or temporal bone where the second and third branches of the fifth pair of cranial nerves emerge from the cranial cavity.

Anteriorly the sphenoid has a foramen for the passage of the Eustachian tube, the *tuba auditiva,* and also a sharp-pointed projection, the nasal portion, called the *rostrum.*

The Frontal. *Location.*—The frontal bones, or ossa frontales, two in number, form the forehead, a portion of the nose, and a portion of the roof of the orbital cavities. They are related posteriorly with the parietal, laterally with the temporal and zygomatic, and anteriorly with the nasal and premaxillary. They touch inferiorly the ethmoid.

Description.—Each of these bones has a *processus orbitalis* which is seen at the outer margin of the posterior and upper orbital wall and just in front of the zygomatic process of the squamous portion of the temporal. The frontal bone forms the anterior portion of the superior wall of the cranial cavity. The two bones are thin, flat, and meet at the median line of the forehead. The external surface is convex. The inner surface has a ridge located longitudinally and in the center the bone becomes narrow anteriorly.

The Parietal. *Location.*—The parietal bones, or ossa parietalia, two in number, form the posterior part of the roof of the cranial cavity. They meet in the median line and are related posteriorly with the occipital, anteriorly with the frontal, and laterally with the temporal bones.

Description.—The parietal bones are short and very broad (Fig. 5, Part II, No. 8). Each bone is quadrilateral in outline and has two surfaces, four borders and four angles. The external parietal surface is convex and smooth and the internal, cerebral surface is concave.

The Temporal. *Location.*—The temporal bones, or ossa temporales, two in number, are located at the lateral portion of the cranium, and aid in the formation of the cranial wall. The temporal bone is related posteriorly with the occipital, superiorly with the parietal and frontal, externally with the quadratus, anteriorly with the ethmoid, and inferiorly with the sphenoid.

Description.—The temporal bones consist of the fused squamous and petrous temporals; they contain the essential organs of hearing.

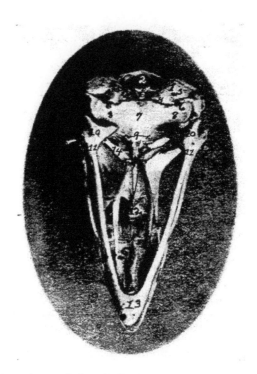

FIG. 8.—Inferior view of the skull of a hen. 1, Occipital. 2, Foramen magnum. 3, Occipital condyle and just below the basi-occipital. 4, Foramen for the passage of the hypoglossal nerve. 5, Foramen for the passage of the vagus. 6, Canalis caroticus and jugulare. 7, Sphenoid. 8, Temporal wings of the sphenoid. 9, Foramen auditiva. 10, Os articulare. 11, Os angulare. 12, Vomer. 13, Os dentale. 14, Pterygoid.

The *squamous portion* of the temporal bone possesses the long thin zygomatic process sometimes called the *posterior orbitalis*. It forms a small flattened tongue, directed forward, sometimes free, and at other times united by its superior border to the summit of the orbital process. This is especially true in the turkey. This process is seen near the lower outer portion of the posterior orbital wall. The squamous portion is also provided antero-laterally with an articular facet which articulates with the quadrate bone. The large portion of the temporal bone lies on the side of the cranium superior to the ala sphenoida temporale. It extends outward and anteriorly over the rims of the petrosum and ala sphenoida orbitale. The squamous part forms the upper three-fourths of the cochlea, the inner auditory canal, the upper part of the fenestra ovalis, the anterior vertical and the outer semicircular canal, and the lower part of the posterior vertical semicircular canal.

The *petrous portion* forms the posterior wall of the foramen ovale and the fenestra ovalis in which lies the columella. The fenestra ovalis and the fenestra rotunda are the only two entrances into the labyrinth. In this region may be seen the fusion line between the os petrosum and the os occipitale. Superiorly and posteriorly the petrous portion touches the external parietalia and occipitalis; infero-laterally it unites with the basi-sphenoid. The outer rim of the foramen ovale is broadened by the ala sphenoida and mesially by the basi-sphenoid. This foramen gives exit to the second and third divisions of the fifth pair of cranial nerves.

The lateral surface of the temporal bone presents a short tube, the *external acoustic process*, or *processus acusticus externus*, to which is attached the concha of the ear. The process is directed outward. Its lumen, the *external acoustic meatus*, or *meatus acusticus externus*, conducts to the cavity of the middle ear in the bare skull, but is separated from it by the tympanic membrane in the natural state.

BONES OF THE FACE

The bones of the face, or ossa faciei, are the premaxillary, or upper mandible, the maxillæ, the nasal, the lacrimal, the palatine, the pterygoid, the zygomatic, the vomeral, the jugal and the quadrato-jugal. The vomer is single, the others are paired. In the inferior maxillary group there are the quadrati, and the inferior maxillary, or lower mandible. The turbinated and hyoid bones are also discussed with the bones of the face.

The beak consists of the anterior portion of the upper and lower mandibles which are covered with a horny structure.

The Premaxilla. *Location.*—The premaxillæ, or ossa incisiva, or mandibular structures, are located in the extreme anterior facial region.

Description.—The premaxillæ are long and the anterior end is pointed. Each consist of two lateral halves which become fused before the chick is hatched. They partly circumscribe the openings into the nose. This bone is the base of the upper portion of the beak and determines its form. It forms the anterior walls of the nasal cavity. It has between the nasal bones, two processes which extend back to the anterior point of the cerebral cavity. The posterior part of the incisivum and nasale are flat, thin, and elastic. The extensions then are the *processus maxillaris*, the *processus palatinus*, and the *processus frontalis*, the first of which forms part of the jaw rim, the second, which aids in forming the gum plate, and the last which reaches to the anterior portion of the cranial wall.

The Maxilla. *Location.*—The superior maxillaries, or ossa maxillares, two in number, form the floor of the upper beak, a part of the palatine roof and nasal walls.

Each maxilla borders laterally the premaxilla and the nasal; inferiorly, the anterior point of the palatine; and posteriorly, the jugale.

Description.—The maxillæ are thin, flat, bones. They have three borders and three angles. The palatine processes of the two bones do not meet in the median line, which results in a cleft in the median palatine region. The bone-like gums are formed partly by the palatine processes but more largely by lateral wings of the os incisivum. Each maxilla has posteriorly a yoke-like extension, superiorly a small extension, and also a palatine process. The latter articulates by a facet with the vomer.

The Nasal. *Location.*—The nasal bones, or ossa nasalia, two in number, are located in the lateral facial region. The nasal bone articulates posteriorly with the frontal; laterally, on the inner border, with the processus frontalis of the premaxilla, and on the outer border with the lacrimal; and inferiorly with the premaxilla, maxilla, jugale, and vomer.

Description.—The nasal bone, or os nasale, is broad, flat, and elastic and forms the posterior wall of the nasal opening. Under this bone is located the infraorbital sinus.

There are three extensions: first, the *processus intermaxillaris*, which forms the upper wall of the nasal cavity; second, the *processus maxillaris anterior*, directed downward and forward, which fuses with the maxillary bone and forms the posterior rim of the nasal cavity; and third, a *posterior processus frontalis*, which lies parallel with the ethmoid.

The Lacrimal. *Location.*—The lacrimal bones, or ossa lacrimalia, two in number, are located at the outer border and at the junction of the processus frontalis of the nasal, with the frontal bone.

Description.—The lacrimal bone is small, and rather filiform. They become fused with the nasal and the frontal bones, forming part of the margin of each.

The Palatine.—*Location.* The palatine bones, or ossa palatina, two in number, enter into the formation of the inner part of the bony gum and hard palate, or roof of the mouth. They form the support for the hard palate. Each palatine articulates posteriorly with the rostrum, or nasal portion of the sphenoid, and with the pterygoid; and anteriorly with the maxilla.

Description.—The palatine bone is curled posteriorly and is thin at the upper portion and thick at the lower border. Anteriorly it has a long rather filiform projection.

The Pterygoid. *Location.*—The pterygoid bones, or ossa pterygoidea, two in number, are located back of the region of the posterior nares. They extend diagonally outward and backward from the median region of the sphenoid rostrum to the quadrate bone. They articulate anteriorly with the sphenoid rostrum and the palatine, and posteriorly with the quadrate.

Description.—The pterygoid bones are slender and cylindrical, and are expanded at the ends into an articular facet.

The Zygomatic. *Location.*—The zygomatic, or ossa zygomatica, two in number are situated below the orbital cavity and extend from the maxilla to the quadrate bone.

Description.—The zygomatic is small, slender, rod-shaped, and forms the lateral portion of the upper jaw. The anterior portion of the zygomatic represents the jugal and is fused with the maxilla and with the anterior processus maxillaris of the nasal bone, the maxilla, and the posterior portion, the quadrato-jugal, which articulates with the quadrate bone.

The Vomer. *Location.*—The vomer is located in the median nasal septum. It articulates with the rostrum of the sphenoid, being connected to it by a mass of ligaments. It touches anteriorly the posterior portion of the maxilla.

Description.—The vomer is a median bone and aids in the formation of the septum nasi. It consists of a thin plate, thickest posteriorly and diminishing toward the anterior edge.

The Jugal and Quadrato-jugal.—The jugal and quadrato-jugal are united forming a long slender cylindrical bone called the zygomatic, lying at the outer side of the upper jaw. They have been described under zygomatic, which see.

The Quadrate. *Location.*—The quadrate bones, or ossa quadrata, two in number, are located antero-laterally to the temporal bones. Each articulates inferiorly with the posterior articular portion, or pars articulare, of the inferior maxilla and infero-laterally with the quadrato-jugalare portion of the zygomatic. It articulates antero-internally with the pterygoid and supero-posteriorly with the temporal.

Description.—The quadrate bone is anvil-like in shape. It has an anterior process, the orbital process, for muscular attachments, and posteriorly it affords attachment to the ear drum.

The Inferior Maxilla. *Location.*—The inferior maxilla, lower jaw, or os maxillare inferius, also called the mandibular bone, or pars dentis, is analogous to the lower jaw of mammals. It articulates posteriorly with the quadrate bone.

Description.—The inferior maxilla is the largest bone of the face. It is made up of a *right* and a *left limb* which are separate in the fetus and which unite subsequently anteriorly, forming the inferior portion of the beak. Each limb of the jaw is developed from five elements: the *pars articularis*, which forms the jaw-joint and, expanded, articulates with the quadrate bone; the *pars angularis*, lying just in front of the pars articularis; the *pars supra-angularis*, a slender bone lying just above the angularis; the *splenial*, a thin plate of bone, lying along the inner surface of the mandible; and the *pars dentalis*, which forms the anterior portion of the jaw.

The Turbinate Bones. *Location.*—The turbinate bones, or ossa turbinata, six in number, are attached to the lateral walls of the nasal cavity (Fig. 26, No. A, 1 and 2). In each nasal cavity there are three turbinate bones, one anterior and two posterior. Of the two posterior the upper one lies superoposterior to the inferior one. The turbinate bones are attached to the lateral nasal walls, project into the cavity and thus greatly diminishing its extent.

Description.—Each turbinate bone is composed of a very thin lamina, finely cribriform in many places, and in the fresh state, covered on both sides with mucous membrane. These bones are curled and partly membrano-cartilaginous structures which give greater surface in the nasal passage for mucous membrane in which the olfactory nerve terminal filaments are distributed.

The Nasal Cavity.—The nasal cavity, or cavum nasi, is a longitudinal passage which extends through the upper part of the face. It is divided into right and left halves by a median septum nasi. Its walls are made up of the premaxilla, maxilla, nasal, vomer, and palatine bones.

The Hyoid Bone. *Location.*—The hyoid bone, or os hyoideum, is situated chiefly between the rami of the mandible, but its upper part extends around the outer margin of the base of the skull. The hyoid bone does not articulate with any bones of the skull, but is firmly attached by strong fibrous structure.

Description.—The hyoid bone consists of the following parts: the *body*, or *basi-hyal*, which is subcylindrical and presents in front a trochlear articular surface, convex transversely, and concave vertically for articulation with the ewer-shaped portion of the glosso-hyal. The anterior free portion, or lingual process, is called the *glosso-hyal*, or *entoglossal bone*. The lingual process gives support to the muscular and fibrous structures of the tongue. On either side of the basi-hyal there is a limb passing posteriorly along the side of the superior larynx, extending upward along the outer border of the occipital bone. This is the *cornu* of the os hyoideum and is divided into two elements, first, the *basi-branchial* which is bone and articulates with the basi-hyal, and the second, the *cerato-branchial*, cartilaginous in structure. In the center and projecting backward from the body of the os hyoideum is a spur process called the *uro-hyal*, partly bony and partly cartilaginous, and which rests upon the superior larynx.

THE VERTEBRAL COLUMN

The vertebral, or spinal column consists of 42 bones, as shown in the following table:

Cervical region	14
Dorsal region	7
Lumbo-sacral region	14
Coccygeal region	7
Total	42

Many of the bones of the dorsal and lumbo-sacral regions do not have free articulations. The cervical and coccygeal alone have free movements.

The Cervical Vertebræ (Fig. 4, No. 59; Fig. 5, Part II, No. 2). *Location.*— The cervical vertebræ form the neck of the fowl.

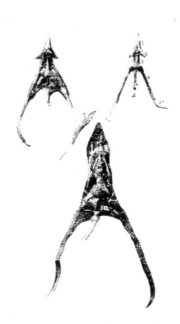

FIG. 9.—Os hyoidum and tongue muscles. 1, Glosso-hyal. 3, Basi-hyal. 4, Basi-branchial element of the cornua. 5, Cerato-branchial element of the cornua. 6, Uro-hyal or spur process. 7, Genio-hyoideus. 8, Cerato-hyoideus showing the slender tendon. 9, Hyoideus transversus. 10, Hyoideus transversus.

Description.—The long cervical section of the spinal column is S-shaped and is made up of fourteen vertebræ. The anterior segments move freely forward, the middle ones backward and the posterior ones forward, allowing the S-shaped curve of the neck. The neck is flexible so that it is possible for the beak to reach the coccygeal oil gland. The bird is enabled to reach the feathers on all parts of the body to cleanse and oil them.

The Atlas. *Location.*—The first cervical vertebra is called the atlas. Anteriorly it articulates with the single condyle of the occipital bone. Posteriorly it articulates with the axis, or second cervical vertebra.

Description.—The atlas is the smallest of the cervical vertebræ and is ring-shaped. The anterior articular surface, half-moon in shape, forms a deep articular cavity. The joint, called a ball-and-socket joint, makes possible movements in all directions. The condyle of the occiput also touches an articular end of the odontoid process of the axis, or second cervical vertebra. Posteriorly there projects from the atlas two small lateral wings possessing articular facets which articulate with similar facets on the lateral wings of the axis. Posteriorly the body of the atlas is also provided with an articular surface which articulates with a similar surface on the anterior portion of the body of the axis.

The Axis. *Location.*—The axis is the second cervical vertebra. It articulates anteriorly with the atlas, and by a facet on the extreme anterior end of the odontoid process, with the condyle of the occipital bone. Posteriorly it articulates with the third cervical vertebra.

Description.—The body of the axis is short. The upper anterior portion of the body of the axis is provided with a tooth-like process called the *odontoid process*. There are two anterior lateral wings provided with small articular facets which articulate with similar facets of the atlas. The anterior surface of the body of the axis forms a true articulation with the body of the atlas.

The axis is provided with a *superior* and an *inferior spine*. There are two posterior articular processes which articulate with the prezygapophyses, or anterior articular processes, of the third cervical vertebra. The posterior part of the body of the axis forms a true articulation with the body of the third cervical vertebra.

Other Cervical Vertebræ.—Beginning with the axis, the body of which is relatively short, the body of each succeeding vertebra is longer than that of the preceding. The articulations of each vertebra with adjoining vertebræ are effected by means of diarthrodial facets, convex in one direction and concave in the other.

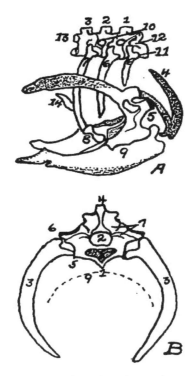

FIG. 10.—*A*. Diagram of three first dorsal vertebræ and scapular arch-side view. *B*. Diagram of section through the hemal arch.

A. 1, First dorsal segment. 2, Second. 3, Third. 4 and 5, Hemal arches. 6, The pleurapophyses or floating ribs. 7, The third pleurapophyses or dorsal vertebral rib articulating with the hemapophysis or sternal rib, 8, and this in turn with the wing of the sternum form a hemal arch. 9, The sternum. 10, Intervertebral foramen through which the nerves pass. 11, Articulation with the body of vertebra parapophyses. 12, Articulation with transverse process or diapophyses. 13, Oblique process or zygapophysis. 14, The epipleural appendage.

B. 1, Hemal or inferior spine (hypapophysis). 2, The neural arch. 3, The pleurapophyses or rib. 4, The superior or neural spine or neurapophysis which aids in the formation of the anapophyseal ridge. 5, Parapophyseal surface for head of rib. 6, Tubercle of rib articulating with the diapophysis. 7, Lamina of neurapophysis. 8, Centrum. 9, Hemal arch.

Between the bodies of the vertebræ are pads of fibrous cartilage. Above these bodies and inferior to the neural spines extends throughout the spinal column the large *neural canal* which is occupied by the spinal cord. Between the vertebral segments the neural canal is exposed superiorly, since here the neural arches do not completely bridge the canal. These spaces are protected by intervertebral ligaments.

In addition to the superior neural spines and the inferior spines, from the body of the vertebra project *lateral processes*, and anterior and posterior *oblique processes*.

All cervical vertebræ except the atlas and the axis are made up of the following parts: a body, or centrum, a neural canal, a neural arch, a superior dorsal, or neural spine, or neurapophysis, which in most cervical vertebræ is only slightly developed, two oblique processes, or zygapophyses, two transverse processes, or diapophyses (Fig. 10, No. 12) and in some vertebræ two tubercles, or anapophyses above the posterior zygapophyses. The *prezygapophyses* are directed upward and inward; the *postzygapophyses* are directed downward and outward. The inferior spines are only well developed in the last two or three cervical vertebræ.

Between vertebral segments, except the central sacral portion, on each side, there is an *intervertebral foramen* (Fig. 11, No. *A*, 4), through which pass the spinal nerves. In the central portion of the sacrum where the vertebræ have fused, there are two foramina on each side for each original vertebra, one located above the other, the upper giving passage to the sensory branch, the lower to the motor branch of the nerve.

The Dorsal Vertebræ (Fig. 4, No. 23; Fig. 11, *A*). *Location.*—The dorsal, or thoracic vertebræ, or vertebræ thoracicales, aid in forming the roof of the chest cavity.

Description.—The dorsal vertebral section is made up of seven vertebræ, with strong short bodies. The first and sixth dorsal segments articulate as do the cervical, by the bodies and oblique processes. (Fig. 11, No. *A*, 3, illustrates the oblique processes.) The seventh dorsal is fused with the first lumbo-sacral vertebra. The second to the fifth inclusive of the vertebræ are fused together, and the superior and inferior spinous processes are fused into a prominent plate-like ridge.

The transverse processes of the dorsal vertebræ, from the second to the sixth, are well developed and are bridged over with a thin layer of bone. The ventral spines are partly fused and form a very prominent and continuous ridge. (Fig. 11, No. *A*, 1, shows the fused superior spines; No. 2, the fused inferior spines; and No. 4, the intervertebral foramina.)

The Lumbo-sacral Vertebræ (Fig. 5, Part II, No. 4). *Location.*—The fused lumbo-sacral section of the spinal column forms the roof of the pelvic cavity.

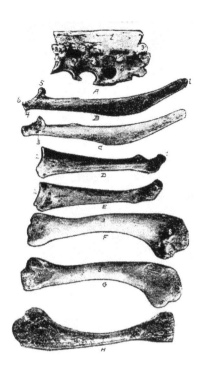

FIG. 11.—Bones from the scapular arch.

A. Dorsal vertebra. 1, Superior spinous ridge, 2, Inferior spinous ridge. 3, Oblique processes. 4, Intervertebral foramina. 5, Articular facette for head of the rib. 6, Articular facette for tubercle of the rib. 7, Articular portion of the body.

B. Outer surface of scapula.

C. Inner surface of scapula. 1, Thin caudal end. 2, Articular head. 4, Processus furcularis. 5, Processus humeralis. 6, Processus coracoideus.

D. Outer surface of coracoid.

E. Inner surface of coracoid. 1, The furcular tuberosity. 2, Articular facette for articulation with the sternum. 3, Articular surface for humerus and scapula.

F. Outer surface of humerus and G the inner surface of the same. 1, Trochanteric fossa. 2, Oval articular head at proximal end. 3, Shaft. 4, Distal end showing articular condyles. 5, Trochanter. 6, Trochlea for ulna. 7, Trochlea for radius. 8, Depression or fossa.

H. Section through the median plane of the humerus showing the delicate cross partitions illustrating provisions made for bones provided with air-sac extensions.

Description.—The lumbo-sacral region consists of fourteen vertebræ which are distinct in the body of the newly hatched chick, but which become fused soon after hatching. With these vertebral segments there are fused the last dorsal and first coccygeal vertebral segments. This fusion is so complete that the segments are indicated only by the intervertebral foramina on the sides through which the spinal nerves pass, and by transverse markings on the inferior surface of the bodies.

The lumbo-sacral vertebræ, called the sacrum, and the ilia are fused. The dorsal spines of the vertebral segments are indicated only in the anterior portion where they are fused, forming a plate. There are no prominent ventral spines.

The Coccygeal Vertebræ (Fig. 4, No. 27). *Location.*—The coccygeal or caudal vertebræ, or vertebræ coccygeæ, constitute the bones of the tail.

Description.—There are seven coccygeal vertebræ. The last segment, shaped like a plough share and therefore called the pygostyle, is the largest and is supposed to have been formed by the fusion of several original distinct segments. It supports the coccygeal oil gland and the row of rudder feathers, or rectrices, which are arranged fan-wise.

All of the coccygeal vertebræ except the first are freely movable, thus allowing the tail to be used as a rudder during flight. The lateral spines are long and well developed, and the superior spines are bifurcated, thus giving increased surface for muscular attachment. The first coccygeal segment is fused with the last lumbo-sacral vertebra.

THE THORAX

The dorsal vertebræ superiorly, the ribs laterally, and the sternum, or breastbone inferiorly, form the skeleton of a large cavity called the thorax. The dorsal vertebræ have been described.

The Ribs (Fig. 4, No. 32; Fig. 10, *A* and *B*). *Location.*—The ribs form the lateral bony wall of the thorax, articulating superiorly with the dorsal vertebræ.

Description.—The ribs are arranged in order of length, the ultimate rib being the longest. From anterior to posterior, they approach more nearly a horizontal position. The ribs are divided into the true and the false. The *true ribs* articulate with the sternum. The *false ribs* do not touch the sternum; they are floating. The true ribs are composed of two parts, a vertebral, or dorsal, and a sternal, or ventral. The vertebral part, or *extremitas vertebralis*, is provided with an articular head, or *caput costæ*; a neck, or *collum costæ*; and an articular tubercle, or *tuberculum costæ*. The head and tubercle articulate with the dorsal vertebra. Below the head of each rib is a *pneumatic foramen*.

The distal extremity of the dorsal section of the rib articulates with the proximal end of the sternal section by a diarthrodial articulation.

The articulations of the true ribs with the sternum is diarthrodial and each articulate by two small ridges with a double sternal facet. The first, the second, and the seventh ribs are floating, or false ribs. The first rib articulates with the quadrate part of the last cervical vertebra and also with the first dorsal vertebra. The seventh, the last rib, articulates with the under side of the anterior alar part of the ilium. This rib is situated similarly to the true ribs; but, instead of articulating directly with the sternum, the lower end lies against the sternal segment of the rib just anterior to it.

From the posterior edge of the second, the third, the fourth, and the fifth ribs, and near the middle of the dorsal segment, are flat, uncinate, bony processes which project upward and backward, overlying in each case, the succeeding rib and giving greater surface for muscular attachments and greater stability to the thorax.

The Sternum (Fig. 4, No. 40; Fig. 10, No. *A*, 9). *Location.*—The sternum, or breast-bone forms the inferior portion, or floor of the thoracic cavity.

Description.—The sternum is a quadrilateral, curved plate with processes projecting from each angle and from the middle of the anterior and posterior borders. The posterior medial projection, or *metasternum*, is the longest, and has a tall, plate-like ridge—the *sternal crest, crista sterni*—running along its ventral surface. The crest serves the important function of increasing the bony area for the attachment of the powerful muscles which move the wings. The anterior medial projection, or *rostrum*, is short, and pierced at its root by an opening from which extend two elongated, saddle-shaped depressions into which the end of the coracoid bones are received.

The plate-like process of bone, the *posterior lateral process*, which projects from the caudal angles of the sternum soon divides into two parts. The shorter of these, the *oblique process*, broadens toward its free end and covers the sternal segments of the last two ribs. The sides of the sternum are thus provided with an external and an internal process forming an *external* and an *internal* notch. These notches are bridged over by a broad ligament, to which the muscles are attached. In poor flyers, as the domestic fowl, these notches are large. The posterior end of the sternum is called the *xiphoid process* or *processus xiphoideus*. Anteriorly the sternum is provided with lateral external processes, the *costal processes*. The lateral borders of the sternum are pitted by four depressions into which the sternal segments of the ribs are received. The dorsal, or inner surface of the bone is pierced by openings by which the air-sacs communicate with the interior.

The Appendicular Skeleton

The appendicular skeleton consists of the shoulder girdle, the fore limb, the pelvic girdle, or hip bone, and the hind limb. The shoulder girdle consists of the scapula, the coracoid, and the clavicle. The fore limb consists of the arm, forearm, and the hand. The arm consists of the humerus, the forearm of the radius and the ulna; and the hand consists of the carpus, the metacarpus, and the phalanges. The pelvic girdle consists of the ilium, the ischium, and the pubis. The hind limb consists of the thigh, the leg, and the foot. The thigh consists of the femur, and the leg consists of the tibia and the fibula. The foot consists of the metatarsus and the phalanges.

THE SHOULDER GIRDLE

The shoulder girdle consists of the scapula, the clavicle and the coracoid.

The Scapula. *Location.*—The scapula (Fig. 11, *A*, *B*) lies on the outer and superior rib surface, extending parallel with the dorsal vertebræ.

Description.—The scapula is a thin, sword-like bone, becoming thicker as it approaches the shoulder-joint. The scapula expands and becomes thin near the free end, which reaches nearly to the antero-lateral portion of the ilium (Fig. 4, No. 24). The scapula articulates with the coracoid. A *pneumatic foramen* is located at the base of the acromion process. The anterior part of the scapula is provided with an articular head and is provided with an inner process, called the *processus furcularis*, which lies near the furcula and coracoid. An outer stronger *processus humeralis* forms the posterior half of the *glenoid cavity*, or *cavitas glenoidalis*, in which the humerus articulates and the processus coracoideus.

The Clavicle. *Location.*—The clavicles, commonly known as the wish bone, are located in the anterior chest region.

Description.—The clavicles are long, slender bones uniting below in the *hypocledium*, a laterally flattened process. They are joined to the upper end of the coracoid by fibrous cartilage. The hypocledium is joined to the anterior point of the sternum by the claviculosternal ligament. The clavicles, as united by the hypocledium, form a v-shaped structure called the furculum, or, popularly, the wish bone.

The forks play the part of an elastic spring, whose office it is to prevent the wings from coming toward each other during contraction of the depressor muscles. The conformation of this bone is, therefore, like the sternum, related to the extent and power of flight; and for this reason it is that, in swift flyers, the two branches of the furculum are thick, solid, widely separated, and curved like a U; while in those that fly heavily and with difficulty, these branches are thin and weak, and join at an acute angle. The latter formation greatly diminishes its strength, and lessens, in a singular manner, the reactionary power of the bony arch it represents.

The Coracoid. *Location.*—The coracoid is located just back of the clavicle and at the side of the entrance of the thoracic cavity. It is the strongest bone of the shoulder girdle, extending upward, outward, and forward. It articulates inferiorly with the sternum and superiorly with the humerus and the scapula, and is attached to the superior end of the furcular limb by a fibrous cartilage.

Description.—It is thinnest in the center or shaft and broadens toward the inferior extremity. The upper hook-like part of the coracoid forms the fore part of the glenoid cavity, and together with the scapula and furcula form the foramen triosseum through which passes the tendon of the elevator muscle of the wing. The upper end flattens out into three tuberosities, the tuberositas furcularis which is thick and to which attaches the limb of the furcula, the tuberositas scapularis which unites to the scapula, and the tuberositas humeralis which lies between these and articulates with the humerus.

THE FORE LIMB

The bones of the fore limb are humerus, ulna, radius, carpus, metacarpus, and phalanges.

The Humerus. *Location.*—The humerus constitutes the arm, which, when at rest, lies parallel to the dorsal vertebræ. It articulates superiorly with the glenoid cavity, a shallow ball-and-socket joint, formed by the scapula and the coracoid; and inferiorly with the ulna and radius.

Description.—The proximal extremity of the humerus is provided with a trochanter (Fig. 11, *G*, No. 5) and a large oval head or caput humeri, which articulates in the glenoid cavity. The head is an elongate, semi-oval convexity, with the long axis transverse from the radial to the ulnar side and with the ends continued into upper and lower crests. The upper crest of the head of the humerus is on the radial side and the lower crest on the under side. Under this latter crest there is a *pneumatic fossa* (Fig. 11, *F*, No. 1), at the upper end of which there is an opening into the bone, the *pneumatic foramen*, which brings the air-sac into communication with the air space of the bone. The shaft, or corpus humeris, is irregularly cylindrical and slightly S-shaped.

The proximal part of the shaft, which is expanded on the palmar side, is concave across and convex lengthwise. The distal part is slightly flattened. The shaft of the humerus is almost cylindrical.

The distal extremity of the humerus is provided with two articular condyles, one of which articulates with the ulna and the other with the radius. On the radial side of the palmar surface there is a ridge; and on the ulnar side of the same surface there is a second ridge diverging to the opposite tuberosity. The radial surface is a narrow, subelongate convexity, extending from the middle, approximately, of the palmar surface, obliquely to the lower part of the radial tuberosity. The two *articular convexities*, or *trochlea*, at the distal end of the humerus are bent toward the palmar aspect, the *anconal aspect* is the side on which the elbow is situated. The inner convexity is the larger and articulates with the ulna. To the outside is the *processus cubitalis humeri*. The outer articulates with the radius and is so arranged that the radius makes a greater flexion than the ulna. At the lower end of the humerus there is a depression which receives the anconeus of the ulna during flexion and extension of the forearm. On the shoulder- and elbow-joint are found grooves over which the tendons glide, at which places sheaths are provided. This arrangement also aids in keeping the capsular ligament in place.

The Ulna. *Location.*—The ulna, larger than the radius, is bent and articulates with the radius only at the ends. The two bones are bound together by a ligamentous band. It also articulates superiorly with the distal end of the humerus, and inferiorly with the carpus. When the wing is folded the radius is superior and a trifle to the inner side of the ulna.

Description.—The proximal end is most expanded, and is obliquely truncate for the articular excavation adapted to the ulnar tubercle of the humerus. A short angular process behind the cavity represents the *olecranon*, which is the *processus olecranalis coracoideus ulnæ* to which the ligamentum capsulare cubiti attaches. On the inner side of the head is the internal tubercle of the ulna and on the external side the external tubercle of the ulna. An excavation is noted on the radial side of the proximal end for the lateral articulation of the head of the radius. On the upper side near the articular part is located the *sigmoid cavity*.

The shaft, or corpus ulnæ, decreases in size near the distal end. It is slightly curved, flattened laterally with an internal and an external ridge.

The distal end of the ulna is slightly expanded into a *trochlea* which is sharply convex and articulates with two free carpal bones, the scapho-lunar, or os carpi radiale, and the cuneiform, or os carpi ulnare. The scapho-lunar is placed on the radial side, and the carpiulnare on the ulnar side. The distal extremity of the ulna is provided with a *styloid process* and on the dorsal side with a *tubercle*.

The Radius. *Location.*—The radius lies beside the ulna with which it articulates at each extremity. At the inferior extremity the articulation is rotary. It also articulates superiorly with the humerus and inferiorly with the os carpi radiale.

Description.—The radius, cylindrical in shape, is more slender than the ulna.

The proximal end is expanded, subelliptical, with a concavity for the oblique tubercle and a thickened convex border for articulation with the ulna. This end is provided with the *tuberositas radii*.

The shaft, or corpus radii, is slender, subcompressed, and has a slight bend upward from the ulna. A nutrient foramen occurs in this shaft.

The distal end is expanded and rather flattened with two grooves on the anconal side for passage of tendons. For articulation with the scapho-lunar the radius is provided with a terminal *transverse convexity* produced palmad, which also articulates with the ulna laterally. There is a tuberosity on the radial side of the expansion and inferiorly the *inferior tuberculum ossis carpi radialis* and superiorly a *superior tuberculum ossis carpi radialis*.

The Hand.—The hand is made up of the carpus, the metacarpus, and the phalanges.

The bones of the hand are so arranged as to allow abduction and adduction, or flexion in the ulno-radial plane, movements which are necessary in the outspreading and folding of the wing. Thus the hand of the fowl moves in a state of pronation, without the power of rotation. The carpal bones are so placed between the anterior arm and metacarpus as to reduce the abduction which is necessary to hold and extend the wing, so that the hand or wing be in a fixed position.

The Carpus.—The carpus in the domestic fowl is represented by two bones, ossa carpi, called the *scapho-lunar*, or *os carpi radiale*, and the *cuneiform*, or *os carpi ulnare*. The scapho-lunar is the smaller and is located between the radius and metacarpus. The cuneiform is the larger and is located between the ulna, the radius, and the metacarpus. The cuneiform is somewhat anvil-like in shape, being provided with a body and two prongs.

The Metacarpus.—The metacarpal bones, or ossa metacarpi, two in number, are separated at their middle portion, and consolidated at their extremities. The upper proximal base of the metacarpus is provided with a tubercle, the tuberculum muscularis, and externally the tuberculum ulnare ossis metacarpi. The distal end of the metacarpal bone is provided with a tuberculum articulare.

The Phalanges.—The first three fingers only are represented, and these are rudimentary.

The first finger called the pollex or thumb, consists of but one joint. It is located on the proximal and outer end of the metacarpal bone. It has on its proximal end a tubercle, the tuberculum articulare.

The second finger, the best developed, consists of two phalanxes. These are the main bones extending from the metacarpus. Each articular end is provided with a *tuberculum articulare*.

The third finger, small, cylindrical in shape, is located at the distal and inner side of the metacarpal bone.

THE PELVIC GIRDLE (Fig. 4, No. 25; Fig. 5, Part II, 6)

The pelvic girdle is made up of three bones as follows: the ilium, the ischium, and the pubis, all of which are fused in adult life. They are separate in the baby chick (Fig. 5, Part II, 6). The pelvis together with the lumbo-sacral vertebræ forms a thin, irregular, shell-like structure extending superiorly from the tail to the thoracic region. The sacrum is broad posteriorly and together with the ilium forms the pelvic roof. The ilium and ischium are its lateral walls.

The top surface of the pelvis shows the *fovea ilio-lumbalis dorsalis*, which is bounded mesially by the crista ilii. Between it and the spina lumbalis there is a broad furrow, the bottom of which is formed by the dorsal surface of the lumbar vertebræ. The *sulcus ilio-lumbalis dorsalis* is formed by the rims of the ilia, so that a ridge is observable. The *cavum ilio-lumbale dorsale* is formed by the iliac rims on either side. Anteriorly is the *canalis ilio-lumbalis*, which is formed by the ilium and the lumbar vertebral spines. It is located longitudinally to the vertebræ. The anterior opening of this canal is the larger. Posterior to the *acetabulum* is the *post-acetabular* ridge.

The under part of the pelvis presents three distinct regions. The cavity is divided into the *fovea lumbalis*, or anterior part, the *fovea ischiadica*, or mesial part, and the *fovea pudendalis*, or posterior part. Posterior to these is the region called the *planum coccygeum*.

The fovea lumbalis contains the anterior lobe of the kidney, and is circumscribed anteriorly by the last rib-carrying vertebra, and posteriorly by the transverse process of the last lumbar vertebra.

In the fovea ischiadica which follows, lie the nerve plexus ischiadicus and the middle lobe of the kidney. Its posterior boundary is the linea arcuata. The linea arcuata is a line drawn from the acetabulum of the one side to the acetabulum of the other side.

The fovea pudendalis contains the posterior lobe of the kidney. The posterior boundary is the anterior border of the first coccygeal vertebra. It gives passage to the nerve plexus pudendo-hemorrhoidalis.

The ilium, the ischium, and the pubis join to form the cotyloid cavity, or acetabulum, in which articulates the head of the femur. The floor of the cavity is perforated by a relatively large round foramen.

The Ilium. *Location.*—The ilium, together with the lumbo-sacral vertebræ, forms the roof of the pelvic cavity. It articulates at its inner border with the lumbo-sacral vertebræ, postero-laterally with the ischium and at the cotyloid cavity with the pubis.

Description.—The ilium is remarkable for its development in the long axis of the vertebral column. It is long and narrow; and mesially, where it is thickest it forms the upper wall of the acetabulum. Anterior to the acetabulum it is outwardly concave, and posterior to the acetabulum it is convex. The ilium fuses with the last dorsal vertebra and with the lumbo-sacral vertebræ and is excavated on its internal face. This surface is irregular and lodges the kidneys. This inner margin of the renal part enters with the square extensions of the posterior excavation passing into the posterior iliac spine. The external margin is the extension of the crista transversa and forms the *processus ischiadicus*. Posteriorward the renal part of the ilium joins the ischium.

The ilia converge at the summits of the anterior sacral spines forming the *ilio-lumbar* spines.

The ilium is joined to the square extensions of the last sacral vertebra by the *symphysis ilio-sacralis*, to the larger part of the lumbar vertebræ by the *sutura ilio-lumbalis*, and to the transverse processes of the last sacral vertebræ by the *symphysis ilio-sacri*.

The Ischium. *Location.*—The ischium is located in the postero-inferior part of the pelvis. It joins superiorly with the ilium, and inferiorly with the pubis.

Description.—The ischium is smaller than the ilium and is a flattened, triangular-shaped bone, thickest where it forms the posterior part of the acetabulum, becoming thinner and broader as it extends backward. Posteriorly it forms the caudal extension. The inferior border is turned slightly outward and is fused with the pubis. Between these bones there is located the large oval *ischiadic foramen*, through which passes the ischiadic nerve. This bone aids the ilium in the formation of the obturator foramen through which passes the tendon of the internal obturator muscle. The lower part of the ischium which separates the ischiadic foramen from the obturator foramen is called the *ramus ascendens ossis ischii*.

The Pubis. *Location.*—The pubis is located along the inferior margin of the ischium and extends further back than the posterior border of the ischium.

Description.—The pubis is a long, slender, rib-like bone. It forms the lower and front portion of the acetabulum. The ischiadic foramen formed by the ischium and pubis is single and nearly circular in fowls, double in pigeons, and in geese and ducks it is elongate.

THE HIND LIMB (Fig. 4, Nos. 35, 47, 48, 49, 50, 51, 52, and 53; Fig. 12)

The pelvic, or hind, limb supports the body. The bones of each leg are the femur, which constitutes the thigh, the tibia and the fibula, which represents the leg, and the metatarsus and the phalanges, which form the foot.

The Femur (Fig. 12, *A* and *B*). *Location.*—The femur, or os femoris, the first bone of the hind extremity, extends obliquely downward and forward, articulating with the acetabulum above and the tibia, the fibula, and the patella below.

Description.—The femur, one of the largest, thickest, and strongest bones of the body, belongs to the class of cylindrical bones, and presents for examination a shaft and two extremities.

The superior, or proximal extremity, is provided with a head, caput femoris; the neck, collum femoris; and the trochanter. The head is relatively small, and is marked by a depression above for the round ligament which fills the cavity in the acetabular wall. Its axis is nearly at right angles with the shaft. A neck joins it to the body at the proximal end. External to the head of the femur is the *trochanter*. The trochanter presents an outer convexity over which the tendon of the gluteus maximus extends to become inserted below. The *trochanteric ridge*, which is opposite the articular head, presents an outer flattened surface which possesses impressions for muscular attachments. The *trochanteric fossa*, or fossa trochanterica, is shallow.

The *shaft*, or *corpus femoris*, is shorter than the tibia, is in general cylindrical, bent forward, and the lower half is flattened and expanded transversely. A nutrient foramen is located in the median portion of the shaft. The shaft shows muscular linear ridges.

The distal extremity is large in both directions and comprises the *trochlea* in front and two *condyles* behind, one internal and one external. The condyles are separated by the *intercondyloid fossa*, or *fossa intercondyloidea*, which is marked with pits for the attachment of ligaments; and above this is the *epitrochlear fossa*.

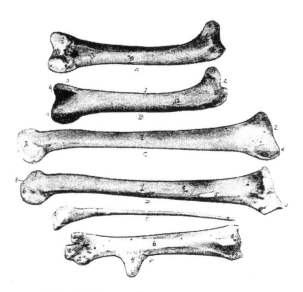

FIG. 12.—Bones of the hind extremity.

A. Posterior view of femur.

B. Anterior view. 1, Shaft. 2, Proximal extremity. 3, Distal extremity. 4, Articular head. 5, Trochanter major. 6, Shallow trochanteric fossa. 7, Convexity over which the tendon of the gluteus maximus glides. 8, External condyle. 9, Internal condyle. 10, Nutrient foramen. 11, Intercondyloid fossa. 12, Muscular linear ridges. 13, Epitrochlear fossa.

C. Internal view of tibia.

D. External view. 1, Shaft. 2, Articular head. 3, Distal end. 4, The rotular process. 5, The fibular ridge. 6, External condyle. 7, Internal condyle. 8, Intercondyloid space. 9, Nutrient foramen.

E. Fibula, lateral view. 1, Articular head. 2, Its attenuated portion.

F. Posterior view of metatarsus. 1, Trochlea for inner or second digit. 2, Trochlea for middle or third toe. 3, Trochlea for outer digit. 4, Bony core for spur. 5, Bony canal for tendons. 6, Groove for tendons. 7, Fossa intercondyloidea.

The *inner condyle* begins anteriorly as a ridge, and expands into a convexity which attains its greatest breadth posteriorly where it becomes more flattened. The inner side of the inner condyle is flattened and is provided with a tuberosity at its mid-part and a second just above the posterior part of the condyle.

The *outer condyle* is formed in the same manner as the inner condyle. It is indented at its broad, lower end by an *angular groove*, which, winding divides the posterior part of the condyle into two convexities. The more external convex ridge and the groove dividing it from the outer condyle are adapted to the head of the fibula. There is in this part a *fibular ridge* and above this ridge a tuberosity.

The Tibia. *Location.*—The tibia (Fig. 12, *D* and *C*) extends obliquely downward and backward from the knee-joint to the hock. It articulates above with the femur and, by its procnemial process with the patella, below with the metatarsus, and laterally with the fibula.

Description.—The tibia is the longest bone in the posterior limb, and possesses a shaft and two extremities. It is largest at the proximal end and presents three faces.

The proximal extremity presents a semi-oval articular surface, not quite at right angles with the shaft, which articulates with the condyles of the femur. The margin is raised toward the anterior of the bone. The *head of the tibia*, or *caput tibiæ*, extends into a rotular process which extends transversely, and is truncate. From the anterior of this process there descends two vertical ridges; one near the angle of the rotular process, the *procnemial ridge*; the other from the outer fibular angle, the *ectocnemial ridge*. On the outer side of the intercondylar tuberosity there is a surface for the ligamentous union with the head of the fibula; and a short distance below this there is a vertical ridge for the close attachment, almost a fusion, with the fibula, called the *fibular ridge*.

The *shaft*, or *corpus tibiæ* is straight and the upper two-thirds subtrihedral; the lower third oval. A *nutrient foramen* occurs near the upper postero-internal portion of the middle third of the bone.

The distal extremity is much smaller than the proximal one; it is quadrangular in form. The expanded inferior end of the tibia forms two articular condyles above which posteriorly there is the *epicondyloid fossa*. The inner condyle, the larger, has a groove near the lower end of the anterior part of the shaft, which deepens toward the intercondyloid space. This *intercondyloid fossa* in young birds is covered by a strong ligament, which in older birds, becomes ossified. On the lateral side of each condyle, there is a depression for the attachment of ligaments.

The Patella. *Location.*—The patella, or knee-cap, thin and wide, articulates with the procnemial process of the tibia, and with the deep trochlea of the distal end of the femur.

Description.—The patella is irregular in shape with three faces and three borders. The posterior surface is articular. The other two surfaces are rough for tendinous attachment. The patellar ligaments in old birds may become ossified.

In order to turn the foot in and out, the tibia not only turns around the inner condyle of the femur, but also around the patella, so that the posterior surface turns outward and thus turns outward the metatarsus and all the toes.

The Fibula. *Location.*—The fibula lies at the outer border of the tibia. It articulates superiorly with the outer condyle of the femur, and laterally with the tibia.

Description.—The fibula is rudimentary; it is largest superiorly and tapers to a slender point. The head is compressed laterally, and furnishes an upper and an inner articular surface.

The Tarsus.—There exist, during fetal development of the chick, two rows of tarsal bones which later become fused. The upper row fuses with the tibia and the lower with the metatarsus. Therefore, in the adult, there is no tarsus.

The Metatarsus. *Location.*—The metatarsal bone extends downward and forward. In birds it consists of one bone, which articulates superiorly with the inferior extremity of the tibia. On the distal end it has a threefold trochlear arrangement which articulates with the three principal digits.

Description.—The proximal extremity posteriorly has a process which may be considered as a consolidation of originally separate metatarsal bones. The process at the supero-posterior part of the metatarsus is called the *hypotarsus* of the *tarso-metatarsus*, through which extends a canal called the *hypotarsal canal*, and which gives passage to flexor tendons.

The proximal end of the metatarsus is ossified from one center of ossification forming an epiphysis which caps the ends of the three original metatarsal bones that coalesce, first with one another, then with the epiphysis, thus forming a single compound bone. Above and just to the inside of the metatarsus there may occur a small bone which is imbedded in ligaments and articulates with the inner proximal surface of the metatarsus. This has been called by Gadow a *sesamoid bone*.

The shaft shows *tendonal grooves* which are best marked on the posterior surface. The shaft is rounded at the sides and flattened on the anterior and posterior surfaces. At the juncture of the middle and inferior thirds of the metatarsus there is a conical process turned slightly backward, which serves as a base for the *spur*. The spur is a horny structure.

At the distal extremity occurs the threefold trochlear arrangement mentioned above, which incloses the fossa intercondyloidea. The inner trochlea is the broadest, and the outer the narrowest. The inner trochlea articulates with the proximal end of the second, the middle with the third, and the outer with the fourth toe.

The Phalanges. *Location.*—Most domestic fowls are provided with four digits, or toes; the Houdan and Dorking are provided with five. In fowls with four toes the three principal toes, the second, the third, and the fourth, are directed forward, and the first, or hallux, is directed backward.

Description.—The last phalanx of each toe, called the ungual phalanx, is slightly curved downward, is pointed anteriorly, like the claw of the cat or dog, and is covered with a horny sheath. The articular ends of the joints of the phalanges are enlarged. The base of the basal phalanx has two enlargements, the superior-inferior tubercles, or tuberculum superius et inferius, between which is located the fossa articularis transversa. Laterally the head has two condyles, condyli laterales, which are divided by the sulcus longitudinalis. The bodies of the phalanges vary in form; they are superiorly rounded, but inferiorly rounded, flat, or even somewhat concave. The distal extremity has an articular trochlea.

The *first toe*, or digit, called the *great toe*, or *hallux*, is composed of three phalanges, or segments. The first segment, or basal phalanx is considered a rudimentary metacarpal bone; it is attached by a fibrous cartilaginous tissue to the inner posterior surface of the inferior extremity of the metacarpal bone just below the spur.

The *second toe*, likewise composed of three phalanges, is directed forward. It articulates with the inner trochlea located on the inferior metacarpal bone.

The *third toe*, made up of four phalanges, is the middle of the three forward toes. It articulates with the middle trochlea of the inferior extremity of the metacarpal bone.

The *fourth toe*, is composed of five phalanges; it is, however, of approximately the same length as the third, the segments being shorter. The fourth or outer toe articulates with the outer trochlea of the distal extremity of the metacarpal bone.

ARTHROLOGY

Kinds of Joints.—Joints may be movable, immovable, or mixed. In *movable*, or *true joints* the articular surface of each bone is covered by cartilage. The bones are held together by ligaments, the capsular one often surrounding the joint and enclosing the synovial membrane. In some joints there is a pad of fibrous cartilage interposed between the two articular cartilages. Such a pad, called a *meniscus*, adds to the elasticity and the free movement of the joint. Movable joints form the most numerous class; they are for the most part found in the limbs.

In an *immovable joint* there is only a thin layer of fibrous or cartilaginous material interposed between the bones. The fibrous layer of the periosteum of both bones unite to cover the connecting material and becomes attached to the same, thus serving as a ligament. If the connecting material is fibrous, the joint is called a *suture*; if cartilaginous, a *synchondrosis*. These joints are found in the skull and in the pelvis.

The term *mixed* is used with reference, not to the motion in joints, but to their structure, which partakes of the nature of both the movable and immovable. The bones are firmly joined by a strong interposed pad of fibrous cartilage to which also is adherent the ligaments of the joint. There are no capsular ligaments; the cartilaginous pad or disc is softer toward its center, where occasionally there may be one, or even two, narrow cavities. Authorities differ as to whether such cavities are lined by synovial membrane or not. Since there are really no frictional surfaces in such a joint, motion depends upon the flexibility of the disc. The joints between the vertebral centra afford the best illustration of the mixed class.

Movement of Joints.—The movements admissible in joints may be divided into four kinds: gliding, angular movement, circumduction, and rotation. These movements are often, however, more or less combined in the various joints. It is seldom that there occurs only one kind of motion in any particular joint.

Gliding movement is the most simple kind of motion that can take place in a joint, one surface gliding or moving over another without any angular or rotary movement.

Angular movement occurs only between the long bones. By it the angle between the two bones is increased or diminished. It may take place in four directions: forward and backward, consisting of flexion and extension; or inward and outward from the medial line of the body, consisting of adduction and abduction. Abduction of a limb is movement away from the medial line of the body. Adduction of a limb is movement toward the medial line of the body.

Circumduction is that limited degree of motion which takes place between the head of a bone and its articular cavity, whilst the extremity and sides of the limb are made to circumscribe a conical space, the base of which corresponds with the inferior extremity of the limb, the apex with the articular cavity; this kind of motion is best seen in the shoulder- and hip-joint.

Rotation is the movement of a bone upon an axis, which is the axis of the pivot on which the bone turns, as in the articulation between the atlas and axis, when the odontoid process serves as a pivot around which the atlas turns; or else is the axis of a pivotlike process which turns within a ring, as in the rotation of the radius upon the humerus.

Pronation is a form of rotation in which the inferior extremity of the radius passes before the ulna, and thus causes the hand to execute a kind of rotation from without inward.

Supination is a form of rotation in which the movement of the forearm and hand are carried outward so that the anterior surface of the latter becomes superior.

The Ligamentous Structure.—Ligaments are dense, fibrous, connecting structures. They are made up principally of white fibrous tissue and exist in all true joints.

There are four kinds of ligaments associated with true joints:

The first kind, the *capsular ligament* which encloses all true joints, is thin and consists of interlaced fibers attached to the bone at the edges of the articular cartilages. It either partly or wholly surrounds the joint, enclosing and protecting a synovial apparatus, which, by secreting a liquid resembling serum, lubricates the joint to prevent friction.

The second kind, the binding or *lateral ligaments*, consist of flattened or rounded cords or bands of fibrous tissue. Such a ligament extends from one bone to the other, and firmly attached to their roughened surfaces, holds the bones in place and at the same time allows the required amount of motion.

The third kind, located between the joints, is called *interosseous ligament.*

The fourth kind, called the *annular ligament* binds down and protects the tendons.

Ligaments of the Ear[2].—The concha of the ear is provided with a superior and an inferior ligament.

2. The classification of Gadow is used.

Ligaments of the Jaw.—The articulations of the lower jaw are complex. The freely movable articulation is between the inferior maxilla and the quadrate. Less freely movable articulations are formed by the quadrate with the temporal, the zygomatic and the pterygoid. A lateral ligament of the jaw, the *articulo-jugale* extends from the posterior border of the inner wing of the os articulare of the inferior maxilla (Fig. 8, No. 10, and Fig. 19, No. 15) to the outer border of the os quadrato jugulare.

A *lateral temporo-maxillary ligament* extends from the outer surface of the temporal bone to the outer border of the os articulare of the inferior maxilla (Fig. 8, No. 11, and Fig. 19, No. 14).

Ligaments of the Vertebræ.—In each space between the bodies of the vertebræ there is a *meniscus intervertebralis*. This meniscus is analogous to the annulus fibrosus of mammals (Gadow), which, is possibly formed as a protrusion of the anterior surface of the vertebral body. The meniscus or disc may develop into a ring-shaped structure, the true meniscus, or it may develop to different degrees as an extension of the vertebral body surface, and become, as in the dorsal and lumbar regions, fused with the vertebral segments, in which case it is called the annulus fibrosus.

The vertebral disc which is connected with the atlas and which is fused with that bone, represents the first meniscus, which is called the *ligamentum transversum atlantis*. This, as well as other ligaments of the spinal column, may become ossified.

The ligament which is located between those vertebral bodies which face each other and is inside the joint cavity, is called the *ligamentum suspensorium corporum vertebralium*. It passes through the central opening of the meniscus and lies exactly in the long axis of the body of the vertebra.

The first ligament of the neck is the *ligamentum suspensorium dentis epistrophei*. The *ligamentum capsulare atlantico-occipitale* and the *ligamentum capsulare atlantico-epistrophicum odontoideum* are two parts of the joint capsule of the vertebral body. Other ligaments of this part are the *membranæ obturatoriæ intervertebrales posteriores*, which are located between the semicircular rims of the neck vertebræ.

The *ligamentum transversum atlantis* surrounds the occipital condyle.

The *ligamentum nuchæ* is a thin, membranous, ribbon-like structure which lies between the muscles of the right and those of the left side of the middle, and the lower part on the posterior of the neck, and ends in attachments to the superior spines of the cervical vertebræ.

The *ligamentum elasticum interspinale profundum* and the *ligamentum elasticum interspinale superficiale* and the three last named are the ligaments which keep the neck of the fowl in the s-shape, without the action of the muscles.

The *ligamentum capsulare obliquum* connects the facets of the oblique processes.

Ligaments of the Ribs.—The thick inferior end of the sternal portion of the true rib has two small articular heads, which articulate with two depressions in the articular surface of the sternum. This articulation is held firm by a *capsular ligament*.

The upper end of the sternal rib forms an almost perfect right angle with the inferior end of the dorsal rib with which it is connected by a joint provided with a synovial apparatus and a capsular ligament. This arrangement allows free movement outward and inward and is the main joint in respiration.

The upper end of the dorsal portion of the rib articulates with the dorsal vertebra. The joint formed by the articular head with the body of the vertebra is provided with a capsular ligament. The tubercle of the rib articulates with the facet on the transverse process of the dorsal vertebra, and is provided with a transverse ligament, called the *ligamentum transversum externum*. It is also provided with a capsular ligament.

The *ligamentum triangulare* connects each processus uncinatus with the succeeding rib. This ligament is in the form of a membranous sheet, or aponeurosis.

Ligaments of the Sternum (Fig. 13).—The external and the internal notches of the sternum are bridged over with a thin membrane which gives an extensive surface for muscular attachments. The sternal muscles overlie this portion.

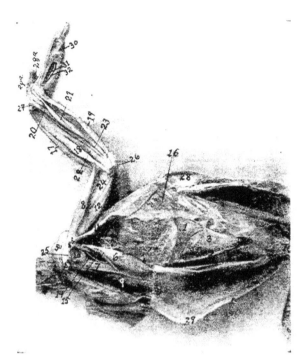

FIG. 13.—Muscles of the fore extremity. Inside view. 1, Lateral external process of sternum. 2, Lateral internal process of same. 3, Ligament of the external notch. 4, Ligament of the internal notch. 5, Pectoralis major. 5*a*, Its fan-shaped expansion at shoulder-joint laid back. 6, Pectoralis tertius. 7, The coracoid. 8, Biceps. 8*a*, Its long head. 8*b*, Its short head. 9, Pectoralis secundus. 10, Rectus abdominis. 11, Teres et infraspinatus. 12, Deltoid. 13, Capsular ligament. 14, Teres minor. 15, Coraco-brachialis. 16, Serratus magnus anticus. 17, Extensor metacarpi radialis longior. 18, Pronator brevis. 19, Flexor carpi ulnaris. 20, Flexor carpi ulnaris brevior. 21, Flexor digitorum profundus. 22, Extensor indicis longus. 23, Extensor ossis metacarpi pollicis. 24, Humerus. 25, Shoulder. 26, Elbow. 27, Carpus. 28, Section through vertebra. 29, Keel of sternum. 28*a*, Flexor brevis pollicis. 29*a*, Extensor proprius pollicis. 30, Flexor minimi digiti brevis. 31, Interosseous palmaris. 32, Interosseous palmaris.

The sternum is connected by a fibrous mass with the inferior portion of the clavicle, or hypocledium (Fig. 57), the *claviculosternal ligament*.

The inferior narrow elongated end of the coracoid forms with the sternum a true articulation, which is provided with a capsular ligament, the *ligamentum capsulare*. Capsular ligaments occur at the articulations between the sternum and ribs.

Ligaments of the Shoulder-joint (Fig. 13).—The shoulder-joint is made up of the scapula, the humerus, and the coracoid. The ends of these three bones form the *foramen triosseum* (Fig. 15, No. *A*, 7) through which passes the tendon of the elevator muscle of the wing.

The furcula, independent of the shoulder-joint or girdle, is attached to the supero-internal part of the proximal end of the coracoid by fibrous connective tissue; it is also connected to the other bones of the shoulder-joint by the *ligamentum coraco-furculare* and the *ligamentum furculo-scapulare*.

The *ligamentum coraco-scapulare* extends from the tuberosity of the furcula to the coracoid and to the processus furcularis of the scapula.

The *ligamentum coracoido-scapulare externum* extends between the external tuberosity of the coracoid, the tuberosity of the scapula, and the humerus.

The *ligamentum coracoideo-scapulare inferius* extends from the coracoid to the inner tubercle of the scapula.

Another long, broad ligament belongs to the episternal apparatus.

The shoulder-joint is provided with a wide, loose, *capsular ligament* (Fig. 13, No. 13). Attached to the humerus are four other ligaments, of which three pass from the anterior end of the coracoid and the fourth from the scapula. The latter are as follows: first, the *supero-anterior ligamentum humero-coracoideum*, which extends from the small tubercle of the humerus to the coracoid bone; second, the *antero-inferior ligamentum humero-coracoideum*, which extends from the humerus to the coracoid; third, the *ligamentum coraco-humerale*, which extends from the coracoid to the large tubercle of the humerus; fourth, the *ligamentum humero-scapulare* which extends between the processus humeralis of the scapula and the head of the humerus.

Ligaments of the Elbow-joint (Fig. 14).—The elbow-joint is made up of the ulna, radius and the humerus. The ligaments of the elbow-joint are as follows: the *ligamentum capsulare cubiti*, or capsular ligament, which extends from the processus cubitalis of the humerus to the processus olecranalis coracoideus of the ulna and to the tuberositas radii of the radius (Fig. 15, No 5).

The *ligamentum laterale cubiti externum* connects the outer humeral distal extremity with the head of the radius (Fig. 15, No. 2).

The *ligamentum laterale cubiti internum* lies between the inner distal extremity of the humerus and the tuberculum internum of the ulna (Fig. 14, No. 9).

The *ligamentum annulare radii* originates on the olecranon, surrounds the head of the radius, and is attached to the tuberculum internum of the ulna (Fig. 15, No. 3).

The *ligamentum cubiti teres* extends from the head of the radius to the upper end of the ulna (Fig. 15, No. 4).

FIG. 14.—Muscles and ligaments of the arm and forearm of a hen.

A. External view. 1, Expansor secundarium. 2, Tensor patagii longus. 3, Tensor patagii brevis. 5, Brachialis anticus. 6, Pronator longus. 7, Supinator brevis. 8, Extensor indicis brevis. 10, Scapulo-humeralis.

B. Internal view. 9, Ligamentum laterale cubiti internum.

The *ligamentum transversum* spreads out between the head of the ulna and the radius, thus uniting the two bones, and limiting supination (Fig. 15, No. 1).

Ligaments of the Carpal Joint.—The carpal joint is made up of the ulna, radius, the two carpal bones and the metacarpus.

There are two strong *ligamenta obliqua carpi ulnaris* which extend from the processus styloideus of the ulna to the tuberculum posterius carpi ulnaris of the os carpi ulnaris.

The *ligamentum posticum ulnare carpi ulnaris* extends from the processus styloideus of the ulna to the os carpi ulnaris.

FIG. 15.—Ligaments of the arm and forearm of a hen.

A. The scapulo-coraco-humeral articulation. 1, Proximal end of the humerus. 2, Articular head of humerus. 3, The coracoid. 4, Proximal end of the clavicle. 5, Tendon of pectoralis tertius. 6, Tendon of pectoralis secundus. 7, Opening or foramen triossium through which the tendon of the elevator of the wing passes. (Pectoralis secundus.) 8, Broken end of scapula.

B. 1, Ligamentum transversum. 2, Ligamentum laterale cubiti externum. 3, Ligamentum annulare radii. 4, Ligamentum cubiti teres. 5, A portion of the ligamentum capsulare cubiti. 6, Distal extremity of humerus. 7, Ulna. 8, Radius.

C. Muscles of the outside surface of the arm. 1, Tensor patagii longus. 2, Extensor digitorum communis. 3, Tensor patagii brevis. 4, Anconeus. 5, Flexor carpi ulnaris. 6, Biceps. 7, Triceps. 8. Deltoid. 9, Brachialis anticus. 10, Flexor metacarpi radialis. 11, Extensor ossis metacarpi pollicis. 12, Extensor metacarpi radialis longior. 13, Extensor indicis longus.

The *ligamentum ulnare carpi radialis* is a short ribbon-like ligament which passes over the above-mentioned ligaments and is attached to the inner surface of the os carpus radialis.

The *ligamentum ulnare carpi radialis internum* is a strong ligament which originates on the inner part of the elbow. It extends to the upper rim of the base of the main digit.

The *ligamentum radiale carpi radialis externum* originates on the outside of the head of the middle digit and inserts to the superior tuberculum carpi radialis of the radius.

The *ligamentum carpi radialis internum*, a short ligament, extends from the inside of the head of the main digit to the inner rim of the carpi radialis.

The *ligamentum carpi interosseum* is located between the carpal bones.

The *ligamentum ulnare metacarpi internum* is spread out between the processus styloideus of the ulna and the tuberositas muscularis on the upper side of the middle finger base at a point where the second and the third metacarpi separate.

The *ligamentum ulnare metacarpi externum* extends from the tubercle of the distal extremity of the ulna, on the dorsal side, to the tubercle on the base of the metacarpus.

Ligamentum radiale metacarpi extends from the head of the inner digit to a point near the tuberculum muscularis of the radius.

The *ligamentum transversum ossis carpi radialis et metacarpi* is located between the inferior tuberculum ossis carpi radialis and the first metacarpal bone.

There is another ligament between the tuberculum superiorus carpi radialis and the first metacarpal bone.

The *ligamentum ossis carpi radialis internum et metacarpi* extends from the inner surface of the carpus radialis to the tuberositas muscularis of the radius.

The *ligamentum ossis carpi ulnaris externum et metacarpi* extends from the ulnar carpal bone to the tuberculum of the metacarpal bone.

The *ligamentum ossis carpi ulnaris internum et metacarpi* extends from the processus uncinatus of the ulnar carpal bone to the tuberositas muscularis of the metacarpal bone.

Ligaments of the Finger.—The *ligamentum pollicare* connects the thumb with the first metacarpal bone.

The *ligamentum anterius ossis metacarpi et primæ phalangis digiti secundi* extends from the tuberculum articulare metacarpi, of the proximal end of the metacarpus, to the tuberculum articulare of the first phalanx of the second or large finger. Similarly attached, are also an internal and a posterior ligament, or *ligamentum internum* and *ligamentum posterius*.

The three phalanges are all provided with capsular ligaments, or ligamenta capsularia, which bind together the several phalanges.

The small or third finger is connected to the metacarpus by a capsular ligament, the *ligamentum capsulare*, and by an interosseous ligament, or *ligamentum interosseum digitorum*, to the first phalanx of the second finger.

Ligaments of the Pelvis.—The obturator foramen, formed by the ilium and the ischium, is covered by a broad membranous ligament (Fig. 4, No. 29).

The oblong foramen, or foramen oblongum, and the foramen ischiadicum, formed by the ischium and the pubis are covered by broad membranous ligaments (Fig. 4, Nos. 29 and 60).

Poupart's ligament or *ligamentum Poupartii*, quite small in fowls, extends from the anterior lower rim of the ilium to the pelvic cavity, and is inserted at the bottom of the acetabulum.

Another broad ligament originates from the posterior rim of the ischium and is attached to the square surface of the first coccygeal vertebra.

The *ilio-pubic ligament* extends from the pubic spine to the last rudimentary rib.

Ligaments of the Hip-joint (Fig. 25, *F*).—The hip-joint is made up of the ilium, ischium, pubis and femur and is a deep ball-and-socket joint. The cavity or acetabulum is called a cotyloid cavity.

The hip-joint is provided with three ligaments.

The *ligamentum capsulare femoris* is attached around the rim of the cotyloid cavity and around the rim of the articular head of the femur (Fig. 25, No. *F*, 4).

The *ligamentum teres*, or *round ligament*, is a very short ligament which closes the hole at the floor of the cotyloid cavity, at which point it is attached. It is also attached to the head of the femur.

The *ligamentum ilio-sacrale* strengthens the hip-joint capsule. It passes from the lower anterior rim of the os ilii, extending over the capsule to the neck of the femur to which it is attached.

The Ligaments of the Knee-joint (Figs. 16 and 17).—The knee-joint is made up of the femur, patella, tibia and fibula, and is provided with the following ligaments:

The *ligamentum extero-laterale genu* is a strong ligament extending from the condylus externus femoris, or external femoral condyle to the outer surface of the head of the fibula, or capitulum fibulæ. From this head there passes inward, a strong ligament to a point between the femur and fibula (Fig. 25, No. *D*, 4).

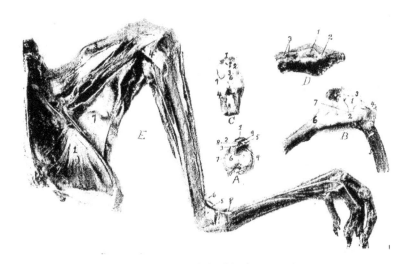

FIG. 16.—Ligaments and muscles of the hind extremity.

A. A transverse section through the tibial or tarsal cartilage, front view. 1, Gastrocnemius tendon. 2, Flexor perforatus annularis primus pedis. 3, Flexor perforans digitorum profundus. 4, Flexor perforatus indicis secundi pedis. 5, Extensor longus hallucis. 6, The articular surface. 7, Ligamentum capsulare ossis tarsi. 8, Flexor perforatus medius secundus pedis. 9, Flexor perforatus indicis secundus pedis. 10, Tendons of the extensor muscles.

B. Knee-joint. Inside view of femoro-tibial articulation. 1, Anterior ligamentum cruciatum. 2, Internal ligamentum laterale genu. 3, Anterior patellar ligament. 4, Patella. 5, Distal end of femur. 6, Proximal end of tibia. 7, Location of meniscus or pad of fibrocartilage.

C. Anterior view of tibio-tarsal articulation. 1, Distal end of tibia. 2, Its articular surface. 3, Ligamentum anticum. 4, Proximal end of metacarpus. 6, Pad of fibrocartilage.

D. The dorsal surface of the coccyx. 1, The bilobate oil gland. 2, Its duct. 3, Levator coccygis.

E. Inside view of pelvis and thigh. 1, Obturator internus. 2, Ambiens. 3, Vastus internus. 4, Internal ligamentum laterale genu. 5, Tibialis anticus. 6, Loop through which the tibialis anticus passes. 7, Adductors of the thigh. 8, Loop for the extensor tendon.

The *ligamentum intero-laterale genu* extends from the outer surface of the internal condyle of the tibia, or condylus internus tibiæ. It gives off a thin ligamentous slip which enters the joint and is attached to the inner half-moon shaped pad of fibrous cartilage, or meniscus, which it draws backward by the flexion of the knee (Fig. 16, No. B, 2).

The *ligamentum popliteum* arises from the fossa poplitea of the distal extremity of the femur and extends downward to the posterior rim of the head of the tibia.

The *anterior ligamentum cruciatum genu* originates from the fossa poplitea and extends outward antero-laterally to the rim of the head of the tibia (Fig. 25, No. 6).

The *posterior ligamentum cruciatum genu* is a short, strong ligament which originates from the cavity of the internal condyle of the femur, or condylus internus femoris, and inserts into the internal glenoid cavity of the tibia (Fig. 25, No. 1) or cavitas glenoidalis interna.

The meniscus of the femoro-tibial articulation is divided into four parts, as follows: first, the *internal adhesio cornu antici cartilaginis lunatæ*; second, the *posterior adhesio cornu antici cartilaginis lunatæ*; third, the *external cornu cartilaginis lunatæ*; fourth, the *anterior cornu antici cartilaginis lunatæ*. The latter originates between the condyles and passes around the ligamentum cruciatum genu posticum (Fig. 16, No. B, 7).

The inner semi-lunar fibrous cartilage is well developed and is joined by two ligaments between the femoral and the tibial surfaces (Fig. 25, No. 1).

The external semi-lunar cartilage lies in a cavity between the external femoral condyle and the head of the fibula (Fig. 25, No. 2).

The fibrous cartilages are bound posteriorly by a ligament, and by a second ligament to the head of the fibula, or capitulum fibulæ. Both pads of fibrous cartilage are connected by a transverse ligament called the *ligamentum transversale commune* (Fig. 25, No. 5).

The patella is provided with several ligaments, as follows:

The *anterior patellar ligament*, broad, strong, and irregular in thickness, extends from the inferior margin of the patella to the rim of the second tibial crest (Fig. 16, No. B, 3).

The *ligamentum capsulare capituli fibulæ* is spread out between the head of the fibula, or capitulum fibulæ, and the superficies glenoidalis peronea tibiæ.

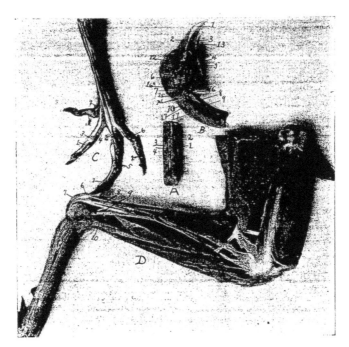

FIG. 17.—Muscles and tendons of the head and posterior extremity.

A. Section of neck with superficial muscles removed. 1, Obliquo-transversales. 2, Interspinales. 3, Intertransversales. 4, Interarticulares.

B. Head showing muscles. 1, Beak. 2, Nostril. 3, Tongue. 4, Sublingual salivary gland. 5, Pterygoideus. 6, Stylo-hyoideus. 7, Temporal. 8, Flexor capitis inferior. 9, Rectus capitis lateralis. 10, Rectus capitis posticus major. 11, Trachelo-mastoideus. 12, Eye. 13, Mylo-hyoideus. 14, Genio-hyoideus. 15, Biventer maxilla. 16, Digastricus. 17, Rectus capitis anterior minor.

C. Foot of hen showing tendons. 1, Flexor perforans digitorum profundus. 2, Flexor perforatus medius primus pedis. 3, Sheath at joint through which tendon passes. 4, Flexor perforatus indicis primus pedis. 5, Flexor perforatus medius secundus pedis. 6, Flexor perforatus annularis primus pedis. 7, Flexor longus hallucis.

D. Outside view of leg of hen. 1, Biceps flexor cruris. 2, Semitendinosus. 3, Loop through which biceps flexor cruris passes. 4, Tibialis anticus. 5, Flexor perforatus indicis secundus pedis. 6, Flexor perforatus medius primus pedis. 7, Gastrocnemius. 8, Flexor perforatus annularis primus pedis. 9, Flexor perforatus medius secundus pedis. 10, Extensor longus hallucis.

The *ligamentum tibio-fibulare* extends from the head of the fibula, or capitulum fibulæ, to the inner surface of the external crest, or crista externa tibiæ, of the tibia (Fig. 25, No. 3).

The *ligamentum interosseum* is a delicate ligament located between the tibia and fibula (Fig. 25, No. 4) which very early becomes ossified.

Ligaments of the Tibio-metatarsal Joint (Fig. 16).—The tibio-tarsal joint is made up of the tibia and metatarsus.

The *ligamentum capsulare ossis tibio-metatarsi* connects the tibia and the metatarsus and surrounds the joint (Fig. 16, Nos. *A*, 7 and *C*, 7).

The *ligamentum externum*, a long strong ligament, originates on the upper and outer surface of the external condyle, or condylus externus of the tibia and is attached to the upper rim of the os metatarsi (Fig. 23, No. 15).

The *ligamentum anticum* is spread out between the fossa intercondyloidea of the tibia and the tuberculum ossis metatarsi of the proximal end of the metatarsus (Fig. 16, No. *C*, 3). Just posterior to this ligament, there is an interosseous ligament, the *ligamentum interosseum*.

A semi-lunar pad of fibrous cartilage, the *cartilago semi-lunaris* is located between the outer distal extremity of the tibia and upper articular surface of the metatarsus (Fig. 16, No. *C*, 6). Its concave portion is directed anteriorly and receives an insertion from the ligamentum externum, or external ligament.

The tendon Achillis (Fig. 16, No. *A*, 1) may become ossified.

Ligaments of the Toes (Fig. 16).—In addition to the capsular ligaments, which all true joints have, we find the following in connection with the toes:

The *ligamentum superius* and the *ligamentum inferius* connect the second toe with the great toe, or hallux.

The *ligamentum laterale externum* and the *ligamentum laterale internum* unite the bases of the second, third and fourth toes to the lateral faces of the trochleas of the inferior extremity of the metatarsus.

The bases of the toes are held together by the *ligamenta transversa*. All the rest of the toes are held together by the *ligamenta capsularia digitorum pedis* and by two lateral ligaments, namely: the *ligamentum laterale externum* and the *ligamentum laterale internum*.

MYOLOGY

Kinds and Structure of Muscles.—Muscles are highly specialized structures which have the property of contractility when stimulated, and thus produce motion. Muscular tissue is sometimes called flesh. Two kinds of muscles are recognized; muscles of locomotion and visceral muscles.

The muscles of locomotion may be in masses of different shape attached to the skin, the *dermal*, or to the skin and skeletal structure, the *dermo-osseous*, or from one bone to another, the *skeletal*. The visceral muscles form sheets and make up a portion of the wall of many of the hollow organs, such as the intestines, the stomach, the gizzard, the esophagus, and the blood-vessels. A special type of muscle forms the heart.

When classified with reference to structure, muscular tissue is divided into three types, as follows: voluntary-striated, involuntary, and involuntary-striated. The microscopic examination of each of these types of muscle shows it to be made up of fibers, these fibers to be made up of muscle cells, the muscle cell to be inclosed in a delicate tubular sheath, or membrane, called the *sarcolemma*. This membrane, tough and elastic, isolates each fiber. The bundles of fibers, called *fasciculi*, are surrounded by a fibrous sheath, which is called the *perimysium internum*; and the entire muscle has likewise an investing sheath of connective tissue, called the *perimysium externum*. The muscular cells show a longitudinal striation marking the fibrillæ, and they also show a cross striation. Nuclei are found just beneath the covering, or sarcolemma, in the striated muscle cell (Fig. 77, No. 2). In the involuntary and in the involuntary-striated muscle cell, the nuclei are centrally located.

The *voluntary-striated muscles* consist of cylindrical fibers and with a few exceptions, are under the control of the will. A muscle of this type, the regular skeletal form, usually has at each extremity a fibrous structure, called a tendon, by means of which it is attached to the bones. The intermediate fleshy portion of the muscle, in case of considerable bulk, is called the belly of the muscle.

The *involuntary, non-striated*, or *smooth muscle*, cells are spindle-shaped, long, and pale in color (Fig. 77, No. 1). The cells lie end to end forming fibers. These fibers do not terminate in tendons, but are arranged in sheets and aid in forming the walls of the digestive tract, to which they give the power of contraction and expansion.

Involuntary-striated, or *heart muscle* (Fig. 77, No. 4), occupies an intermediate position between the two muscles just described. It is composed of cells which branch, are somewhat rectangular. They possess both longitudinal and transverse striation. Among the fibers is found a small amount of connective tissue, as in the former types of muscle, which gives support to the blood-vessels and nerves.

Fascia.—The term fascia is applied to membranous expansions, differing materially in strength, texture, and relations. Fascia is composed of loosely arranged white fibrous connective tissue. At least two layers may usually be distinguished, the superficial fascia and the deep fascia.

Below the skin is the *superficial fascia,* which forms a continuous covering over the whole body and serves to attach the skin to the underlying structure.

The *deep fascia* more densely constructed, may be attached to the skeleton, ligaments, and tendons.

When the fascia spreads out, becomes denser, and acts as a continuation of a muscle, it is called an aponeurosis.

The Muscular Nomenclature.—In the fowl there are 162 muscles, single or in pairs. These muscles are named from their location, as the lingualis; others from their attachments, as the dermo-temporalis; some from their form, as the rhomboideus; others from their use, as the flexor or extensor; and still others from their direction, as the transversus.

THE DERMAL MUSCLES

Birds are provided with a system of delicate muscles divided into numerous fasciculi, which harmoniously act upon the feather quills and collectively agitate the plumage. These are the dermal muscles. This group is divided into two subgroups: the true dermal muscles, that is, those that have their origin and insertion to the under surface of the skin; and the dermo-osseous, those that originate on the surface of some bone and insert to the inner surface of the skin.

TRUE DERMAL MUSCLES

Dermo-frontalis

Dermo-dorsalis

Dermo-tensor patagii

Dermo-humeralis

Dermo-pectoralis

Dermo-frontalis.[3] *Location.*—This muscle is somewhat rudimentary and may be entirely absent. It is located in the frontal region and is about 2 or 3 centimeters long and not so wide.

3. The classification of Shufeldt is used.

Origin and Insertion.—Closely attached to the skin.

Shape.—Flat and rather rectangular.

Relations.—Superiorly with the skin and inferiorly with the frontal bones.

Action.—By contraction the feathers on the top of the head lie flat. Those above the eyes are elevated.

Dermo-dorsalis. *Location.*—In the median line of the neck and back.

Origin and Insertion.—Adhering closely to the skin, it generally becomes lost at the occiput; it is most highly developed in the mid-cervical region, and it gradually disappears over the caudal region.

Shape.—Delicate and ribbon-shaped.

Relations.—Superiorly it is attached to the skin. Fat sometimes surrounds the muscle.

Action.—By contraction it raises the feathers along the superior part of the neck and dorsal region.

Dermo-tensor Patagii. *Location.*—Between the root of the neck and top of the shoulder.

Origin and Insertion.—Attached to the skin in the region of the anterior part of the root of the neck, some fibers passing obliquely upward and blend with fibers of the dermo-temporalis. It blends, by a slender tendon, with that of the tensor patagii longus.

Shape.—A bundle of muscular fibers later becomes thin, delicate, and triangular in shape.

Relations.—Externally to the skin.

Action.—Auxiliary to the tensor patagii longus. A tensor of this region.

The patagii are associated with the wing fold of skin which fills the angle between the arm and forearm. This fold contains elastic tissue and muscle.

Dermo-humeralis. *Location.*—Lateral thoracic region.

Origin and Insertion.—Fan-like delicate fibers from the skin in the abdominal integument, contracting into a long narrow fasciculus of fibers, again spreading out in fan-shape to be inserted to the tendon of the pectoralis major, just below its insertion.

Shape.—Triangular, fan-shape.

Relations.—Superiorly with the skin.

Action.—Controls the skin in this region. The anterior end being fixed, the muscles are raised; and the posterior end acting as the fixed point, the feathers are caused to lie close to the body.

Dermo-pectoralis. *Location.*—On each side of the chest, lying in a longitudinal manner.

Origin and Insertion.—Attached to the skin on either side of the thorax corresponding to the dermo-dorsalis. Anteriorly it disappears over the region of the origin of the cleido-trachealis, and posteriorly just behind the tips of the post-pubic parts of the pelvis.

Shape.—Thin, delicate.

Relations.—Superiorly with the skin.

Action.—The anterior part acting as the fixed point, the feathers of the chest are raised; and the posterior end acting as the fixed point, the feathers are made to lie flat.

THE DERMO-OSSEOUS MUSCLES

Dermo-temporalis

Platysma myoides

Dermo-cleido dorsalis

Cleido-trachealis

Dermo-spinalis

Dermo-iliacus

Dermo-ulnaris

Dermo-temporalis (Fig. 18, No. 1). *Location.*—From the temporal region down the side of the neck to the anterior part of the thoracic region.

Origin.—By a broad tendinous attachment from a small depression above and anterior to the temporal fossa. From here the fibers pass upward and backward and then downward.

Insertion.—A few fibers blend with those of the cleido-trachealis. The fibers are then lost upon the skin in front and opposite the shoulder-joint. Some of the fibers blend with those of the dermo-tensor patagii.

Shape.—Long, flat, and ribbon-shaped. The lower portion consists of delicate fibers.

Relations.—Superior to the temporal muscle and sphenotic process. Superior to the cleido-trachealis and inferiorly to the shoulder-joint. The superior portion touches the skin throughout its entire length.

Action.—A tensor of the lateral cervical integument. It is an auxiliary to the tensor patagii.

Platysma Myoides. *Location.*—The inferior portion of the throat.

Origin.—From the lower margin of the rami of the jaw. Just inferior to the attachment of the masseter.

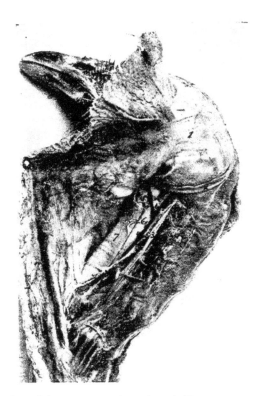

FIG. 18.—Muscles of the upper neck region, 1, Dermo-temporalis. 2, Cornu of the os hyoideum. 3, Carotid artery. 4, Jugular vein. 5, Pneumogastric nerve. 6, Esophagus. 7, Trachea.

Insertion.—From the point of origin it spreads out into a thin fan-like layer and meets its fellow in the median raphe. It is attached to the skin in this region.

Shape.—A very thin fan-shaped muscle.

Relations.—Closely adherent to the skin.

Action.—Assists in supporting the lingual apparatus and superior larynx. Compresses and elevates the part.

Dermo-cleido Dorsalis. *Location.*—Anterior shoulder region.

Origin.—From the upper mesial part of the clavicular bone.

Insertion.—Three fasciculi forming a fan-like arrangement attached to the skin in the shoulder-joint and scapular region. The extremities of the fasciculi may meet in the median line of the dorsal region and merge with the dermo-dorsalis.

Shape.—Three fasciculi forming a fan-shaped radiation.

Relations.—Superiorly with the skin. Inferiorly with the shoulder-joint.

Action.—Contracting with the origin as the fixed point, they brace the skin over the forepart of the back. With the integumental attachment as the fixed point, they aid in the act of respiration.

Cleido-trachealis. *Location.*—Extending from the shoulder to the superior laryngeal region.

Origin.—From a small area on the antero-inner part of the middle of the arm of the furcula.

Insertion.—The fibers pass upward and become flattened and attached to the skin and dermo-temporalis muscle. They touch each other on the anterior border of the superior larynx, the trachea, and the skin over these parts.

Shape.—A slender bundle of muscles.

Relations.—Superiorly the skin, and internally the larynx and the trachea.

Action.—Controls the skin region of this part of the neck.

Dermo-spinalis. *Location.*—In the region of the shoulder.

Origin.—In an attenuated fascia from the crest of the neural spines of the first, the second, and the third dorsal vertebræ.

Insertion.—To the skin in a broad pale stratum, over the scapular region.

Shape.—Thin, pale, and very delicate.

Relations.—Superiorly with the skin, and inferiorly with the shoulder.

Action.—Controls the skin of the region.

Dermo-iliacus. *Location.*—Along the back on either side of the spine.

Origin.—From the inner angle of the marginal portion of the antero-dorsal border of the ilium.

Insertion.—Passing forward as a delicate band it spreads out and becomes lost in the skin of the shoulder region.

Shape.—Thin, delicate, and ribbon-shaped.

Relations.—Superiorly with the skin.

Action.—If the posterior end is fixed, it will cause the feathers to lie close to the skin. Its action would then be opposite to the dermo-dorsalis.

Dermo-ulnaris. *Location.*—Outer surface of the anterior upper costal and the posterior humeral region.

Origin.—By a thin fascia from the outer part of the third and the fourth true ribs at the base of their epineural appendages. Also from the fascia between them.

Insertion.—The fibers pass forward and upward as a thick muscular bundle and is loosely attached to the skin at a point just back of the humerus. The tendon later becomes spread out and attached to the fascia as far as the elbow-joint. It covers the olecranon of the ulna.

Shape.—At first rather thick, fascicular-like; later becomes tendinous; and at its attachment it becomes spread out over the olecranon.

Relation.—Inferiorly with the ribs and humerus, and externally with the skin.

Action.—A depressor of the humeral region.

The Skeletal Muscles
THE MUSCLES OF THE HEAD

Temporal

Masseter

Biventer maxillæ

Entotympanicus

Pterygoideus internus

Pterygoideus externus

Digastricus

Temporal (Fig. 19, No. 4). *Location.*—Occupies the temporal fossa.

Origin.—From the mesian line of the sphenotic process, and the adjacent wall of the orbit.

Insertion.—The fibers passing downward and forward blend with the fibers of the masseter. Inferiorly it inserts by a tendon to the coranoid process upon the superior ramal margin of the mandible.

Shape.—Fan-shaped with broad portion uppermost.

Relations.—It occupies the temporal fossa. It is related superiorly with the dermo-temporalis and skin. Inferiorly, with the biventer maxillæ and the masseter.

Action.—It aids in closing the jaw.

Masseter (Fig. 19, No. 1). *Location.*—Occupies the supero-lateral portion of the surface of the lower jaw.

Origin.—In two portions: the first, by a broad and thin tendon from the entire length of the bony ridge above the external auditory meatus, from the squamosal process, and from the outer portion of the quadrate bone; the second, from the side and under border of the zygoma.

Insertion.—The first portion: The fibers pass downward and forward beneath the zygoma; a few of the fibers blend with those of the temporal; one tendon inserts to a small tubercle on the upper border of the jaw behind the coronoid process, and by fleshy insertion to the outer side of the ramus of the lower jaw, quite as far forward as the horny portion of the beak. The second portion: by a small tendon to the mandible on its upper border immediately in front of the articular portion.

Shape.—Flat, and elongate. Somewhat fleshy.

Relations.—Superiorly with the temporal, and inferiorly with the biventer maxillæ and stylo-hyoideus. Externally with the skin, and internally with the jaw bone.

Action.—Aids in closing the jaw.

Biventer Maxillæ (Fig. 19, No. 3). *Location.*—Covers the outer posterior portion of the mandible.

Origin.—It arises in two portions: The first, from a ridge bounding the posterior part of the auditory canal; the second, from a depression of the mesial side. These two parts blend and extend downward and forward.

Insertion.—To the posterior part of the articular end of the mandible.

Shape.—A curved fleshy mass.

Relations.—Superiorly with the temporal and stylo-hyoideus. Posteriorly with the digastricus and genio-hyoideus. Internally with the jaw bone, and externally with the skin.

Action.—It aids in opening the jaw.

Entotympanic. *Location.*—Posterior to the pterygoid.

Origin.—From the side of the basi-sphenoid and from the base of the rostrum immediately beyond it.

Insertion.—The fibers pass downward and forward, and insert by a double tendinous slip. One slip inserts to a spine-like process on the upper side of the pterygoid, and the other to the quadrato-pterygoidean articulation.

Shape.—Spindle-shaped.

Relations.—Anteriorly with the pterygoid. Internally and posteriorly with the basi-sphenoid. Externally with the pterygoideus internus.

Action.—Aids in raising the upper mandible by pulling forward the quadratus and pterygoidean against the palatines.

Pterygoideus Internus (Fig. 17, No. B, 5). (Synonym.—Pterygoideus medialis.[4])

4. According to international veterinary nomenclature.

Location.—A muscular mass at the roof of the mouth.

Origin.—From the major part of the surface of the palatine bone and the distal head and shaft of the pterygoid and the sphenoidal projection.

Insertion.—By a tendon to the antero-internal part of the articular part of the mandible.

Shape.—Fusiform with a thick fleshy belly.

Relations.—Superiorly with the palatine, the pterygoid, and the sphenoid bones. Inferiorly with the skin.

Action.—It aids in closing the jaw.

Pterygoideus Externus. (Synonym.—Pterygoideus lateralis.)

Location.—Supero-external to the pterygoideus internus.

Origin.—From the outer part of the extremity of the orbital process of the quadrate.

Insertion.—The fibers pass downward, outward and forward and are inserted to the inner part of the mandibular ramus.

Shape.—A small round bundle.

Relations.—Inferiorly with the pterygoideus internus. Externally with the inferior maxilla.

Action.—Aids in closing the jaw.

Digastricus (Fig. 17, No. B, 16). *Location.*—Extends from the basi-temporal to the side of the neck.

Origin.—From the external lateral angle of the basi-temporal.

Insertion.—Opposite the angle of the jaw the fibers spread out in fan-like arrangement. The muscle meets its fellow of the opposite side; extends longitudinally over the superior larynx, for some distance down the neck.

Shape.—Thin and ribbon-shaped.

Relations.—Externally to the platysma myoides. The anterior fasciculi blend with those of the mylo-hyoideus; internally with the larynx at the upper part.

Action.—To raise the trachea and hyoid apparatus against the pharynx.

THE MUSCLES OF THE TONGUE

Mylo-hyoideus

Stylo-hyoideus

Genio-hyoideus

Cerato-hyoideus

Sterno-hyoideus

Depressor-glossus

Cerato-glossus

Hyoideus transversus

Mylo-hyoideus. *Location.*—The forepart of the inter-ramal space.

Origin.—From the inner side of the lower jaw at a point just above the lower border and the inturned edge of the horny sheath of the beak.

Insertion.—By aponeurosis to the under side of the hyoid, on the median line between the first and the second basi-branchial.

Shape.—Thin, flat, and delicate.

Relations.—Inferiorly with the skin. Thin and rather broad, it meets its fellow of the opposite side. Laterally with the rami of the jaw. Superiorly with the hyoid apparatus.

Action.—Lifts the tongue upward against the roof of the mouth.

Stylo-hyoideus (Fig. 17, No. B, 6). *Location.*—Supero-posterior to the hyoid apparatus.

Origin.—From the outer portion of the articular enlargement of the lower jaw.

Insertion.—By a tendon to the basi-branchial of the thyro-hyal.

Shape.—A long transversely flattened fasciculus.

Relations.—Supero-posterior part of the hyoid apparatus along side the genio-hyoideus and the cerato-hyoideus.

Action.—Singly, pulls the tongue to one side; acting with its fellow, pulls the tongue upward.

Genio-hyoideus (Fig. 9, No. 7). *Location.*—Supero-posterior to the hyoid apparatus.

Origin.—From the anterior portion of the inner side of the lower jaw.

Insertion.—To the middle of the outer side of the basi-branchial of the corner of the os hyoides which cornu it completely envelops.

Shape.—A long, rather thick fasciculus.

Relations.—With the stylo-hyoideus and cerato-hyoideus.

Action.—Protrudes the tongue from the mouth.

Cerato-hyoideus (Fig. 9, No. 8). (Synonym.—Kerato-hyoideus.)

Location.—Inferior to the hyoid apparatus.

Origin.—By a delicate tendinous slip from the under side of the shaft of the basi-hyal element of the hyoid apparatus.

Insertion.—To a small elevation on the under side of the anterior end of the glosso-hyal.

Shape.—A small round muscular fasciculus becoming tendinous in its anterior half.

Relations.—Superiorly with the hyoid apparatus.

Action.—Singly, pulls the tongue to one side; acting with its fellow, depresses the tongue.

Sterno-hyoideus. *Location.*—Inferior to the hyoid apparatus.

Origin.—From the outer anterior surface of the thyroid bone of the superior larynx.

Insertion.—To the anterior part of the basi-hyal.

Shape.—At first broad and rather fleshy, becomes contracted and somewhat tendinous.

Relations.—Inferior to the hyoid apparatus and superior larynx.

Action.—When the larynx is fixed, acting alone deflects the tongue laterally; acting with its fellow, depresses the tongue. If the base of the tongue is fixed the two pull the larynx forward. Hence muscles of deglutition.

Depressor-glossus. *Location.*—Superior to the basi-hyal.

Origin.—From the under portion of the basi-hyal.

Insertion.—To under part of the glosso-hyal.

Shape.—Small, short.

Relations.—Superiorly with the basi-hyal and glosso-hyal bones.

Action.—Depresses the tip of the tongue and elevates the base.

Cerato-glossus. *Location.*—On the upper portion of the cornu of the hyoid.

Origin.—One-half of the surface of the first basi-branchial.

Insertion.—To the upper side of the shaft of the cerato-branchial element of the hyoid apparatus.

Shape.—A small muscular fasciculus.

Relations.—Supero-laterally to the cerato-branchial element.

Action.—Elevates the cornua of the hyoid arches and presses them against the skull.

Hyoideus Transversus (Fig. 9, No. 9 and 10). *Location.*—Envelops the spur process of the euro-hyal.

Origin and Insertion.—It extends from the inner side of one cornu to the spur process.

Shape.—Thin, flat.

Relations.—Posterior to the euro-hyal, which it envelops, and between the cornua of the os hyoides.

Action.—Approximates the cornua of the os hyoides.

THE CERVICAL MUSCLES

Complexus

Rectus capitis anticus minor

Flexor capitis inferior

Rectus capitis posticus major

Biventer cervicis

Longus colli posticus

Obliquus colli

Longus colli anterior

Rectus capitis lateralis

Trachelo-mastoideus

Interspinales

Interarticulares

Obliquo-transversales

Intertransversales

Scalenus medius

Complexus (Fig. 19, No. 5). *Location.*—Supero-lateral portion of the neck.

Origin.—By three tendinous slips from the antero-lateral portion of the transverse process of the fourth, the fifth, and the sixth cervical vertebræ.

Insertion.—The muscular fibers pass around the neck and meet their fellows in the median line in a fascia formation. The thin tendinous sheet inserts into the occiput just above the occipital ridge.

Shape.—From point of origin this muscle expands into a broad sheet.

Relations.—It overlies the muscles of the occipital region.

Action.—It extends the head.

Rectus Capitis Anticus Minor (Fig. 19, No. 8). (Synonym.—Rectus capitis ventralis minor.)

Location.—Infero-lateral portion of the neck.

Origin.—From the apices of the hyapophyses of the second, the third, and the fourth cervical vertebræ.

Insertion.—To the occiput just below the complexus.

Shape.—A long and rather thick fasciculus.

Relations.—Externally with the skin. The tendinous slips of origin pass between the flexor capitis inferior and longus colli anterior.

Action.—Extends the head.

Flexor Capitis Inferior (Fig. 19, No. 7). *Location.*—Antero-inferior part of the neck.

Origin.—From the apices of the hyapophyses of the second, the third, and the fourth cervical vertebræ.

Insertion.—To the triangular area on the basi-temporal of the base of the cranium.

Shape.—At first tendinous, becomes somewhat fleshy as it passes forward.

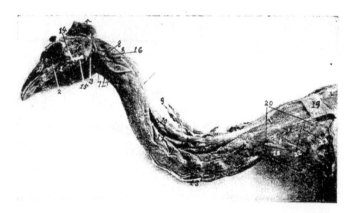

FIG. 19.—Muscles of the head and neck. 1, Masseter muscle. 2, Infraorbital sinus. 3, Biventer maxilla. 4, Temporalis. 5, Complexus. 6, Rectus capitis posticus major. 7, Flexor capitis inferior. 8, Rectus capitis anticus minor. 9, Biventer cervicis. 10, Longus colli posticus. 11, Obliquus colli. 12, Intertransversales. 13, Longus colli anterior. 14, Lateral temporo-maxillary ligament. 15, Lateral ligament of the jaw. 16, Trachelo-mastoideus. 17, Rectus capitis lateralis. 18, Scalenus medius. 19, Rhomboideus. 20, Trapezius. 21, Location of scapula. 22, Supraspinatus. 23, Teres et Infraspinatus.

Relations.—Antero-inferior portion of the neck close to the rectus capitis anticus minor.

Action.—Flexes the head upon the neck.

Rectus Capitis Posticus Major (Fig. 19, No. 6). (Synonym.—Rectus capitis dorsalis major.)

Location.—Supero-lateral area on the anterior portion of the neck.

Origin.—From the median anterior aspect of the second cervical vertebra, and the supero-anterior border of the neural canal.

Insertion.—To the crescent-shaped area at the back of the skull.

Shape.—Short, thick, fleshy bundle.

Relations.—Superiorly with the complexus; inferiorly with the trachelo-mastoideus and the vertebræ.

Action.—Singly, pulls the head to one side; acting with its fellow, extends the head.

Biventer Cervicis (Fig. 19, No. 9). *Location.*—Occupies the superior part of the neck.

Origin.—From the side of the neural spine of the first dorsal vertebra, and from the adjacent tendon of the longus colli posticus.

Insertion.—To the occiput.

Shape.—Tendinous in the middle with a more or less spindle-shaped belly at each end.

Relations.—Superior to the longus colli posticus. It lies next to the skin.

Action.—Extends the head on the neck and elevates the neck.

Longus Colli Posticus (Fig. 19, No. 10). (Synonym.—Longus colli dorsalis.)

Location.—Occupies the superior part of the neck.

Origin.—From the marginal edges of the summits of the neural spines of the first two dorsal vertebræ.

Insertion.—To the transverse process of the axis, and superior part of the cervical vertebræ.

Shape.—A long, somewhat narrow, flattened muscle, the inferior portion of which is divided into five or six fasciculi.

Relations.—Superiorly with the biventer cervicis and inferiorly with the vertebræ.

Action.—Raises the neck upward.

Obliquus Colli (Fig. 19, No. 11). *Location.*—On the lateral side of the neck.

Origin and Insertion.—The first of the seven fasciculi originates from the diapophysis of the eleventh cervical vertebra, winds obliquely over the tenth, and is inserted into the posterior margin of the postzygapophysis of the ninth vertebra. In its passage it receives the slip from the longus colli posticus. The next fasciculus originates from the transverse process of the tenth cervical vertebra, winds obliquely over the ninth vertebra, and is inserted to the postzygapophysis of the eighth vertebra. In its passage it also receives a fasciculus from the longus colli posticus. The next three fasciculi originate in a similar manner being attached to similar postzygapophyses and have slips extending forward which slips are inserted to the neural spines of the alternate vertebræ. The sixth fasciculus originates from the transverse process of the sixth vertebra, passing obliquely up the neck is inserted to the extremity of the diapophysis of the fourth vertebra, the outer extremity of the transverse process of the third vertebra. The seventh, or anterior, fasciculus originates from the transverse process of the fifth vertebra, and is inserted to the extremity of the diapophysis of the third vertebra.

Shape.—Short, thick fasciculi.

Relations.—Laterally with the vertebræ. Inferiorly with the intertransversales. Superiorly with the longus colli posticus, and externally with the skin.

Action.—Flexes one vertebra on the other laterally.

Longus Colli Anterior (Fig. 19, No. 13). (Synonym.—Longus colli ventralis.)

Location.—Occupies the anterior portion of the neck.

Origin and Insertion.—This muscle is divided into a vertical portion, and a superior and an inferior oblique portion. The vertical portion originates from the hypophyses of the tenth to the fifteenth vertebra and is inserted by a tendon to the tubercle on the inferior portion of the atlas. Small, slender tendons are given off and insert to the apices of the parapophyses of the fourth to the tenth cervical vertebræ. There is more or less attachment to the bodies of the cervical vertebræ mentioned. The superior oblique portion originates from the diapophyses of the third, the fourth, the fifth cervical vertebræ and becoming tendinous is inserted to the tubercle on the inferior portion of the body of the atlas. The inferior oblique portion originates from the transverse processes of the fourth, the fifth, and the sixth cervical vertebræ. It is inserted by a slender tendon to the apex of the parapophysis of the third cervical vertebra.

Shape.—A long fleshy muscle extending the entire length of the neck.

Relations.—Inferiorly with the skin and superiorly with the cervical vertebræ.

Action.—Pulls the neck downward.

Rectus Capitis Lateralis (Fig. 19, No. 17). *Location.*—The infero-lateral anterior portion of the neck.

Origin.—Originates tendinous from the diapophysis of the third, the fourth, and the fifth cervical vertebræ.

Insertion.—Passes obliquely upward in front of the spinal column to the inner tubercle on the basal ridge of the basi-temporal bone, by a subcompressed tendon.

Shape.—A somewhat thick fasciculus.

Relations.—Superiorly with the trachelo-mastoideus, inferiorly with the longus colli anterior, and externally with the skin.

Action.—Singly, pulls the head down and to one side; acting with its fellow, pulls the head downward.

Trachelo-mastoideus (Fig. 19, No. 16). (Synonym.—Longissimus capitis et atlantis.)

Location.—Laterally on the anterior portion of the neck.

Origin.—Semitendinous from the diapophyses of the second, the third, the fourth, and the fifth cervical vertebræ.

Insertion.—By subcompressed tendon to the base of the cranium, at the outer tubercle of the basal ridge of the basi-temporal.

Shape.—Flattened from side to side; broad at the posterior portion, becomes angular at the anterior portion.

Action.—Singly, pulls the head down and to one side; acting with its fellow, pulls the head downward.

Interspinales (Fig. 17, *A*, No. 2). *Location.*—Superior to the vertebra.

Origin and Insertion.—A series of muscles connecting the superior neural spines of the cervical vertebræ.

Shape.—Thin, flat.

Relations.—Anteriorly the posterior border of the vertebræ in front and posteriorly the anterior border of the succeeding vertebræ, inferiorly by the vertebræ.

Action.—To approximate the spinous portion of the vertebræ.

Interarticulares (Fig. 17, No. *A*, 4). *Location.*—Supero-laterally to the vertebræ. Between the postzygapophysis.

Origin and Insertion.—From the posterior margin of the ring of the atlas to the postzygapophysis of the axis. Then in the succeeding vertebræ from the postzygapophysis of the vertebra to the same of the succeeding vertebra.

Shape.—Muscular bundle.

Relations.—Inferiorly with the vertebræ. Supero-laterally with the obliquo-transversales.

Action.—Aids in approximating the vertebræ in a supero-lateral direction.

Obliquo-transversales (Fig. 17, *A*, No. 1). *Location.*—Supero-lateral to the vertebræ.

Origin and Insertion.—Passes obliquely from the transverse process of one vertebra to the postzygapophysis of the vertebra beyond.

Shape.—A thin fasciculus.

Relations.—Inferiorly with the interarticulares and laterally with the intertransversales.

Action.—Aids in flexing the vertebræ supero-laterally.

Intertransversales (Fig. 19, No. 12). *Location.*—Laterally to the vertebræ.

Origin and Insertion.—Extend between the transverse processes beginning at the third cervical vertebra.

Shape.—Short, thick.

Relations.—Laterally the vertebræ. Superiorly the obliquo-transversales.

Action.—Aids in flexing the vertebræ laterally.

Scalenus Medius (Fig. 19, No. 18). (Synonym.—Scalenus.)

Location.—Supero-laterally to the entrance of the thorax.

Origin.—From the diapophysis and pleurapophysis of the eleventh cervical vertebra.

Insertion.—To the entire border of the first rib. A few fibers pass over to the anterior free margin of the middle third of the second rib.

Shape.—Rather pyramidal in shape with base uppermost.

Relations.—Posteriorly with the levatores costarum. Internally with the longus colli.

Action.—When the first rib is fixed singly, turns the neck to one side; acting with its fellow, extends the neck. When the neck is fixed, by drawing the first rib forward, it acts as an inspiratory muscle.

THE MUSCLES OF THE AIR PASSAGES

The Superior Larynx

Constrictor glottidis

Thyreo-arytenoideus

The Inferior Larynx

Tracheo-lateralis

Broncho-trachealis posticus

Broncho-trachealis anticus

Broncho-trachealis brevis

Bronchialis posticus

Bronchialis anticus

Sterno-trachealis

Constrictor Glottidis. *Location.*—Supero-anterior portion of the superior larynx.

Origin.—From the superior and longitudinal line of the thyroid plate.

Insertion.—All along the inner margin of the arytenoid bone, and to the apex of the mid-cricoidal segment.

Shape.—Thin, flat sheet extending outward and upward, and then inward.

Relation.—Interiorly with the larynx.

Action.—The two muscles of this kind acting together, close the glottis by drawing the apices of the arytenoids to the median line.

Thyreo-arytenoideus. *Location.*—Supero-lateral portion of the superior larynx.

Origin.—From the entire outer margin of the thyroid plate and the outer margin of the cricoid bone.

Insertion.—All along the outer margin of the arytenoid bone and the outer border of the central cricoid piece.

Shape.—A thin, flat sheet.

Relations.—Infero-laterally with the superior larynx.

Action.—The two muscles of this kind acting together, open the glottis by pulling the arytenoid bone outward.

Tracheo-lateralis (Fig. 20, No. 2). *Location.*—Along the lateral side of the trachea.

Origin.—By the union of the bronchio-trachealis posticus and bronchio-trachealis anticus, on the lateral side of the trachea, about 1½ centimeters from the bifurcation of the trachea.

Insertion.—In delicate fan-like structure to the side of the trachea, near the superior larynx.

Shape.—Thin, ribbon-shaped, closely attached to the trachea.

Relations.—Internally with the trachea, and externally with the skin and other integument.

Action.—Acts as a brace to the sides of the trachea, and contracting, approximates the tracheal rings and thus shortens the trachea.

Broncho-trachealis Posticus (Fig. 21, No. 3). *Location.*—Postero-superior portion of the trachea. The fibers pass downward and backward.

Origin.—From the inferior end of the tracheo-lateralis.

Insertion.—To the end of the third half of the bronchial ring of the same side.

Shape.—A small fasciculus.

Relations.—Internally with the trachea, and externally with the skin and other integument.

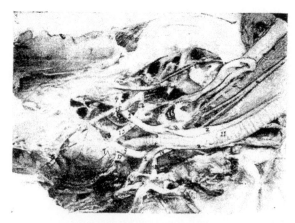

FIG. 20.—The anterior pectoral region. 1, Sterno-trachealis. 2, Tracheo-lateralis. 3, Broncho-trachealis brevis. 4, Costal process of sternum. 5, Right

and left carotid converging and occupying same sheath at 6. 7, The heart showing branches of coronary artery. 8, Cervical air-sac. 9, Left brachio-cephalic artery. 10, Right brachio-cephalic artery. 11, Trachea. 12, Inferior larynx. 13. Subclavian artery. 14, Carotid trunk. 15, Internal thoracic artery. 16 Thyroid gland. 17, Anterior vena cava. 18, Auricular appendage.

FIG. 21.—The inferior tracheal region. 1, Bronchialis anticus. 2, Bronchialis posticus. 3, Broncho-trachealis posticus. 4, Broncho-trachealis anticus.

Action.—A tensor of the true or inferior larynx.

Broncho-trachealis Anticus (Fig. 21, No. 4). *Location.*—Anterior portion of the inferior extremity of the trachea.

Origin.—The anterior branch of the bifurcated tracheo-lateralis.

Insertion.—To the anterior extremity of the third half ring of the bronchus of the same side.

Shape.—A small fasciculus.

Relations.—Internally with the trachea, and externally with the skin and other integument.

Action.—A tensor of the inferior larynx.

Broncho-trachealis Brevis (Fig. 20, No. 3). *Location.*—Posterior part of the inferior extremity of the trachea.

Origin.—From just beneath the broncho-trachealis anticus muscle.

Insertion.—Extends obliquely across the inferior larynx and is inserted to the posterior end of the second bronchial ring.

Shape.—Rather short, strong, straight, subcylindrical.

Relations.—Internally with the trachea, and laterally with the broncho-trachealis posticus.

Action.—A tensor of the inferior larynx.

Bronchialis Posticus (Fig. 21, No. 2). *Location.*—Supero-lateral side of the anterior end of the bronchi.

Origin.—From the lateral inferior margin of the last tracheal ring.

Insertion.—Passing obliquely across the larynx it inserts to the posterior extremity of the second half ring of the bronchus.

Shape.—Small, thick, spindle-shaped.

Relations.—Internally with the trachea, inferiorly with the bronchialis anticus, and superiorly with the broncho-trachealis brevis.

Action.—A tensor of the larynx.

Bronchialis Anticus (Fig. 21, No. 1). *Location.*—Inferior to the former muscle.

Origin.—From the last ring of the trachea.

Insertion.—The fibers pass obliquely forward and become inserted to the rim of the arytenoid cartilage of the inferior larynx, and the anterior extremities of the first and second half rings of the bronchus.

Shape.—Thick, spindle-shaped; about twice the size of the bronchialis posticus.

Relations.—It is crossed at its origin by the broncho-trachealis anticus and is related internally with the bronchus and end of the trachea.

Action.—A tensor of the larynx.

Sterno-trachealis (Fig. 20, No. 1). *Location.*—The inferior region of the trachea above the above-named muscles.

Origin.—From the side of the trachea beneath the posterior border of the broncho-trachealis anticus.

Insertion.—Passes downward and backward across the cavity of the chest and is inserted to the inner part of the costal process of the sternum.

Shape.—A delicate cord of fibers.

Relations.—Internally with the broncho-trachealis muscles.

Action.—A relaxor of the larynx, and hence, of the tympanic membrane.

THE STERNAL GROUP

Triangularis Sterni. *Location.*—In the floor of the thoracic cavity.

Origin.—From the entire superior margin of the summits of the costal processes.

Insertion.—The fibers extend upward and backward, dividing into four digitations which cover the inner surface of the three principal sternal ribs, and are inserted to the first four as high as their articulations with the vertebral ribs.

Shape.—Flat and somewhat triangular.

Relations.—Interiorly with the floor of the thoracic cavity.

Action.—A powerful muscle of expiration. In contracting, lessens the cubic content of the thorax.

The Diaphragm (Fig. 47, No. 10; Fig. 33, No. 9). *Description.*—The diaphragm in the domestic fowl is rudimentary. It consists of a thin semi-transparent membrane, situated between and separating the thoracic and the abdominal cavities. It readily conforms to the various organs pressing upon it from each side. On each side are three rudimentary muscles which extend from the vertebral heads of the second, the third, and the fourth sternal ribs. The fibers of these muscles spread out in fan-like upon the diaphragm, and are seen just above the digitations of the triangularis sterni.

THE ABDOMINAL MUSCLES

Obliquus abdominis externus

Obliquus abdominis internus

Rectus abdominis

Transversalis abdominis

Obliquus Abdominis Externus (Fig. 22, No. 1). *Location.*—The outermost muscle of the lateral abdominal wall.

Attachments.—By a delicate aponeurotic membrane, from the sides of all the true dorsal ribs, from the posterior border of the last vertebral rib and the adjoining margins of the pelvis, and from the entire posterior surface of the inferior border of the post-pubic element of the pelvis; by aponeurosis, blending with the fascia toward the root of the tail and lower part of the abdomen; and by aponeurosis to the sides of the sternum, to the under part of the pectoralis major muscle, and to the zyphoid prolongation. Also by aponeurosis it meets its fellow from the opposite side, at the linea alba.

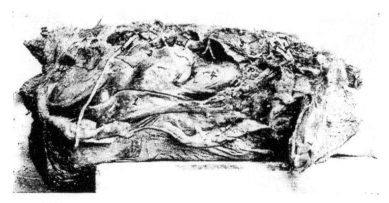

FIG. 22.—The abdominal muscles of a hen. Right side. 1, Obliquus abdominis externus. 2, Obliquus abdominis internus. 3, Peritoneum covering intestines. 4, Peritoneum covering the gizzard. 5, The lungs (note how small they are from a relative standpoint). 6, The great sciatic or ischiadic nerve. 7, Lumbar nerves. 8, Kidney. 9, Brachial plexus. 10, Pericardium of the base of the heart. 11, Anterior vena cava. 12, Oil gland. 13, Showing a sacculation of the abdominal muscles. There is no tunica abdominalis in the fowl.

Obliquus Abdominis Internus (Fig. 22, No. 2). *Location.*—Just internal with regard to the external oblique.

Attachments.—By aponeurosis, from the posterior third of the post-pubic element of the pelvis; by muscular fibers, from the balance of the bone; and by a few fibers, from the iliac border posterior to the acetabulum. Anteriorly, these fibers are inserted to the entire posterior margin of the last vertebral rib, and into the pleurapophysial head of the last costal rib.

Rectus Abdominis (Fig. 13, No. 10 and Fig. 23, No. 1). *Location.*—The inferior median abdominal wall.

Attachments.—By aponeurosis, from the distal extremity of the post-pubic element of the pelvis and from the semitendinous ligament which stretches from one post-pubic tip to the other. Attaches anteriorly to the zyphoid margin of the sternum, and passes, by a broad aponeurotic membrane, over the outer surface of the thoracic wall beneath the external oblique.

Transversalis Abdominis (Fig. 25, No. H, 1). *Location.*—Infero-lateral portion of the abdominal wall.

Attachments.—By a thin, tendinous attachment from the entire post-pubic and iliac margins of the pelvis and from the interpubic ligament. Its fibers cross the abdomen between the peritoneum, the rectum, and the internal oblique. It is inserted over the entire pleural part of the last two vertebral ribs, the intercostal muscles between them, and the same surface of the hemapophyses connected below. Inserts into the linea alba.

Action.—The abdominal muscles give support to the abdominal organs, aid in flexing the spine, draw the last rib backward, thus aiding in respiration, and compress the abdominal organs to aid in defecation and expiration.

THE DORSO-LUMBAR REGION

Sacro-lumbalis

Longissimus dorsi

Sacro-lumbalis. *Location.*—The lateral lumbo-sacral region.

Origin.—Tendinous from the anterior margin of the ilium, from the angles of the last two vertebral ribs, and by tendinous slips, from the outer ends of the transverse processes of the last three dorsal vertebræ.

Insertion.—By a few fleshy fibers into the angle of the first dorsal rib and to the corresponding points upon the free cervical ribs, and by a strong semitendinous insertion into the outer extremity of the diapophysis of the twelfth cervical vertebra.

Shape.—A close fitting, tendo-muscular sheet extending between the anterior margin of the ilium and the root of the neck.

Relations.—Intimately blended with the longissimus dorsi externally.

Action.—Assists the longissimus dorsi.

Longissimus Dorsi (Fig. 24, No. 4). *Location.*—The superior dorso-lumbar region.

Origin.—From the inner portion of the anterior margin of the ilium, and from the various surfaces afforded by its walls and the walls of the ilio-neural canal; by a series of short and distinct tendons alternately from the anterior and posterior extremities of the summits of the neural spines of all the dorsal vertebræ. From the diapophyses of the dorsal vertebræ, from the crests of the neural spines of the last three dorsal vertebræ, from the bodies of the dorsal vertebræ, and from the fascia between them and the sacro-lumbalis; and also by a tendinous sheet continuous with the origin of the longus colli posticus.

Insertion.—By four fasciculi into the free posterior margins of the oblique processes of the eleventh, the twelfth, the thirteenth, and the fourteenth cervical vertebræ.

Shape.—A large flat sheet.

Relations.—Laterally, with the superior spinous processes of the vertebræ, and with the superior surface of the ribs.

Action.—Singly, flexes the back laterally; with its fellow, aids in elevating the body upward.

THE COCCYGEAL MUSCLES

Levator coccygis

Levator caudæ

Transversus peronei

Depressor caudæ

Depressor coccygis

Lateralis caudæ

Lateralis coccygis

Infracoccygis

Levator Coccygis (Fig. 23, No. 2). *Location.*—The superior part of the caudal apparatus.

Origin.—From a limited area of the ilium just beyond and to the side of the anterior free caudal vertebræ.

Insertion.—Into the tuberosity of the anterior margin of the pygostyle.

Shape.—Short, fleshy.

Relations.—Superior portion of the coccygeal vertebræ.

Action.—Elevates the tail.

Levator Caudæ (Fig. 23, No. 4). *Location.*—Supero-lateral side of the tail.

Origin.—From the posterior surface of the post-acetabular area of the pelvis, and from the superior surface of all the coccygeal vertebræ except the pygostyle.

Insertion.—To the four inner quill butts of the main tail feathers.

Shape.—Long, spindle-shaped.

Relations.—Internally, with the levator coccygis and externally, with the lateralis caudæ.

Action.—Powerfully elevates the four inner main tail feathers.

Transversus Peronei (Fig. 23, No. 3). *Location.*—The posterior abdominal region.

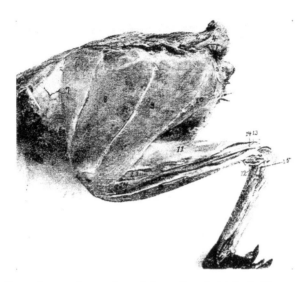

FIG. 23.—Outer layer of muscles of the tail and thigh, 1, Rectus abdominis. 2, Levator coccygis. 3, Transversalis peronie. 4, Levator caudæ. 5, Lateralis caudæ. 6, Depressor caudæ. 7, Sartorius. 8 and 9, Gluteus primus. 10, Semitendinosus. 11, Gastrocnemius. 12, Peroneus longus. 13, Flexor perforatus indicis secundus pedis. 14, Flexor perforatus medius secundus pedis. 15, Lateral ligament of the hock.

Origin.—From the entire posterior margin of the ischium, and from the posterior margin of the post-pubis extending beyond it.

Insertion.—Becoming aponeurotic, it passes toward the coccyx and is attached to the entire posterior margin of the ilium. Passes to the median line, meeting its fellow in front of the anus.

Shape.—Thin, sheet-like.

Relations.—Infero-anteriorly with the anus. Posteriorly with the edges of the ilium and of the ischium. Internally with the depressor caudæ.

Action.—Gives support to the viscera in the post-anal region, and aids in attaching the anus to the structures above.

Depressor Caudæ (Fig. 23, No. 6). *Location.*—The infero-lateral side of the tail.

Origin.—From the lower half of the posterior border of the ischium, and from the entire posterior border of the post-pubis beyond.

Insertion.—To the quill butts of the three or four outer main tail feathers.

Shape.—Strong, conical.

Relations.—Externally, with the depressor coccygis.

Action.—Singly, pulls the tail downward and outward; with its fellow, pulls the tail downward.

Depressor Coccygis. *Location.*—The outermost of the infero-lateral muscles of the tail.

Origin.—From the lower half of the posterior margin of the ischium and the anterior three-fourths of the posterior margin of the post-pubic element of the pelvis beyond it.

Insertion.—To the thickened rim of the inferior and expanded portion of the pygostyle.

Shape.—Flat, triangular.

Relations.—Internally with the depressor caudæ.

Action.—Singly, pulls the tail downward and to one side; with its fellow, pulls the tail downward.

Lateralis Caudæ (Fig. 23, No. 5). *Location.*—The lateral side of the tail.

Origin.—From the tip of the transverse process of the first free caudal vertebræ.

Insertion.—To the outer three quill butts of the main tail feathers.

Shape.—Four fasciculi forming a fleshy belly.

Relations.—Inferiorly, with the levator caudæ and superiorly with the depressor caudæ.

Action.—Singly, pulls the tail downward and outward; with its fellow, pulls the tail downward. The outermost fasciculus contracting, spreads the tail feathers.

Lateralis Coccygis (Fig. 25, No. E, 2). *Location.*—The infero-lateral side of the caudal vertebræ.

Origin.—From the surface of the posterior end of the ilium and by tendons from the under side of the ends of the first three or four caudal vertebræ.

Insertion.—To the side of the posterior margin of the expanded portion of the pygostyle.

Shape.—Subcompressed mass.

Relations.—With the transverse processes of the caudal vertebra, inferiorly.

Action.—Controls the lateral movements of the tail and its feathers, and the oblique downward movement.

Infracoccygis (Fig. 25, No. E, 1). *Location.*—The extreme inferior portion of the caudal vertebræ.

Origin.—From the inferior surface of the diapophysis of the last vertebra which anchyloses with the pelvic sacrum, and from all the free caudal vertebræ.

Insertion.—Into the lower side of the pygostyle.

Shape.—Flat, somewhat triangular.

Relations.—Superiorly, with the vertebræ, laterally, with the lateralis coccygis, and mesially, with its fellow of the opposite side.

Action.—Depresses the tail.

THE COSTAL REGION

Latissimus dorsi

Trapezius

Rhomboideus

Serratus magnus anticus

Serratus parvus anticus

Teres et infraspinatus

Intercostales

Levatores costarum

Appendico-costales

Latissimus Dorsi (Fig. 24, No. 7). *Location.*—Supero-lateral portion of the dorsal region.

Origin.—From two portions, the anterior slip from the outer edge of the superior margins of the neural spines of the second and the third dorsal vertebræ, and the second portion from a similar point on all the succeeding dorsal vertebræ. The second portion is fascia-like in its attachments.

Insertion.—The fibers converge toward the humerus and enter between the deltoid the scapular head of the triceps and the remaining heads of this muscle and is inserted to the anconal part of the humerus, just within in the radial crest.

Shape.—Thin, triangular.

Relations.—Superiorly, with the skin. The most superficial of the dorsal muscles.

Action.—To elevate the humerus and thus flex the shoulder-joint. An expiratory muscle when the wing is fixed.

Trapezius (Fig. 19, No. 20). *Location.*—In the shoulder region, just below the longissimus dorsi.

Origin.—From the neural spines of the second, the third, the fourth, and the fifth cervical vertebræ just below the latissimus dorsi.

Insertion.—To the mesian upper border of the scapula.

Shape.—A flat oblong layer of fibers.

Relations.—The posterior part of the trapezius overlies the anterior part of the rhomboideus. The posterior two-thirds is covered by the latissimus dorsi. The anterior third is superficial.

Action.—Draws the scapula forward.

Rhomboideus (Fig. 19, No. 19). *Location.*—Supero-posterior scapular region.

Origin.—From the neural spines of the first four dorsal vertebræ beneath the latissimus dorsi and trapezius muscles.

Insertion.—The fibers passing outward and backward attach to the posterior third of the mesial or upper border of the scapula.

Shape.—Thin, flat, delicate.

Relations.—Superiorly, by the latissimus dorsi and the trapezius.

Action.—To draw the free end of the scapula upward and forward.

Serratus Magnus Anticus (Fig. 13, No. 16). *Location.*—The supero-lateral side of the thorax.

Origin.—By three strong digitations: first, from the outer part of the second true rib just above the base of the epineural appendage; the other two, from similar locations on the two succeeding ribs, just above the origin of the dermo-ulnaris.

Insertions.—The first by tendinous attachment to the interpleurapophysial membrane. The rest of the fibers pass upward and slightly forward, and are inserted to the inferior surface of the apex of the scapula.

Shape.—Three flat, strong digitations.

Relations.—Internally, with the ribs.

Action.—If the scapula is fixed, it is a muscle of inspiration.

Serratus Parvus Anticus. *Location.*—The outer and upper surface of the anterior part of the thorax.

Origin.—By three digitations from the outer surface of the first three ribs above the origin of the serratus magnus anticus and the thoraco-scapularis.

Insertion.—The thin sheet of fibers passes upward and backward, and is inserted to the inferior margin of the scapula.

Shape.—Thin, flat.

Relations.—Internally, with the ribs.

Action.—If the scapula is fixed, it is a muscle of respiration.

Teres et Infraspinatus (Fig. 13, No. 11). *Location.*—Scapular region.

Origin.—From the superior surface and outer margin of the posterior two-thirds of the scapula.

Insertion.—The fibers pass forward and outward, and insert to the humerus at the middle of the ulnar margin of the pneumatic fossa, between the forks of the triceps.

Shape. —Flat, triangular-shaped.

Relations.—Anteriorly, with the supraspinatus. Internally, with the ribs.

Action.—Aids in keeping the humerus in its socket. Assists in closing the wing by drawing the humerus to the side of the body.

Intercostales (Fig. 24, No. 5). *Location.*—Between the ribs of the upper thoracic region.

Origin and Insertion.—From the anterior border of one rib, the fibers, passing obliquely upward and forward, and the lower portion downward and forward, are inserted to the posterior border of the rib just in front.

Shape.—They are best developed in the anterior portion of the thoracic region and extend down as far as the sternal ribs.

Relations.—Internally the pleura, and anteriorly and posteriorly the ribs.

Action.—The first rib being rendered fixed by the scalenus medius, the muscles draw the ribs forward, thus aiding inspiration.

Levatores Costarum (Fig. 24, No. 2). *Location.*—Supero-lateral portion of the thorax.

Origin and Insertion.—A series of muscles, the first of which extends from the extremity of the transverse process of the twelfth vertebra, the fibers passing downward and backward to the anterior free margin of the upper third of the long posterior rib, and to the external surface close to its margin. The others arise and insert in a similar manner from the ends of the diapophyses of all the dorsal vertebræ and attach to each succeeding rib. Those in front are best developed.

Shape.—Thin and triangular.

Relations.—External to the intercostales.

Action.—Aids in respiration.

Appendico-costales (Fig. 24, No. 1). *Location.*—The lateral side of the thorax.

Origin and Insertion.—From the posterior edge of an uncinate process or epineural appendage, the fibers extending downward and backward to the outer surface of the succeeding rib.

Shape.—A series of thin triangular-shaped muscles.

Relations.—Internally, with the intercostales.

Action.—Aids in respiration.

THE ANTERIOR PECTORAL GROUP

Tensor patagii longus

Tensor patagii brevis

Tensor Patagii Longus (Fig. 15C, No. 1; Fig. 14, No. 2). *Location.*—Anterior shoulder region, in the triangular patagium of the wing.

Origin.—By a flat tendon common to it and the tensor patagii brevis from the supero-mesial line of the head of the clavicle.

Insertion.—By a long slender tendon extending in the duplicature of the patagium, with which the fibers blend. Passing to the side of the extensor metacarpi radialis longus, it extends over the end of the radius and is inserted to the os carpi radiale, and to the fascia which binds down the other tendons on the anterior part of the wrist-joint.

Shape.—Cone-shaped belly terminating in a long tendon.

Relations.—Externally, with the skin; by its tendon, with the border of the extensor metacarpi radialis longior.

Action.—Tenses the soft part in the fold of the wing, and aids in flexing the forearm.

Tensor Patagii Brevis (Fig. 14, No. 3). *Location.*—In the triangular patagium of the wing.

Origin.—From the head of the clavicle in common with the preceding.

Insertion.—The tendon bifurcates, one branch blending with the fascia of the extensor metacarpi radialis longior, the other inserting just below the tubercle on the external condyle of the humerus.

Shape.—Fleshy; somewhat broader and longer than the preceding.

Relations.—Interiorly, with the preceding.

Action.—Assists the tensor patagii longus in flexing the forearm.

THE PECTORAL MUSCLES

Pectoralis major

Pectoralis secundus

Pectoralis tertius

Pectoralis Major (Fig. 13, No. 5). *Location.*—On the lateral side of the sternum.

Origin.—From the posterior portion of the lateral wing of the sternal body, from the outer marginal third of the keel of the sternum, and from the entire outer side of the limb of the furcula or clavicle.

Insertion.—The fibers converging form toward the proximal third of the humerus a broad tendon which by its insertion covers the entire palmar part of the pectoral crest. A few of the tendinous fibers pass over the shoulder-joint and blend with the long head of the biceps. Near this point it receives the insertion of the dermo-humeralis.

Relations.—Interiorly, with the skin; superiorly, in the sternal region, with the pectoralis secundus and the pectoralis tertius.

Shape.—Fleshy; largest of the pectorales.

Action.—Powerfully depresses the humerus. The chief muscle of flight.

Pectoralis Secundus (Fig. 13, No. 9; Fig. 15, No. *A*, 6). *Location.*—The entire lateral side of the sternum.

Origin.—From the anterior sternal extremity of the lower third of the coracoid, from the keel and sternal wing, and by a tendon from a membraneous expansion between the coracoid and the clavicular bones and from the lower third of the coracoid.

Insertion.—The fibers converge into a tendon, which passes upward around the coracoid to its posterior through a canal formed by the scapula, the coracoid, and the clavicle, the foramen triosseum (Fig. 15, No. *A*, 7). This tendon then passes outward and downward, and becoming flat, inserts to the humerus just anterior to the radial crest and nearer to the humeral head than does the pectoralis major.

Relations.—Internally, with the breast-bone; externally, with the pectoralis major; and superiorly, with the pectoralis tertius.

Shape.—Long, fusiform, fleshy.

Action.—Raises the wing.

Pectoralis Tertius (Fig. 13, No. 6; Fig. 15, No. *A*, 5). *Location.*—On the antero-lateral side of the sternum.

Origin.—From the anterior half of the exterior of the body of the sternum, from the fascia of the subclavius on the outer border of the costal process, and from the outer, lower third of the coracoid process.

Insertion.—The fibers pass upward to the outer side of the coracoid, and becoming tendinous as they reach the humerus, by a strong flattened tendon insert to the ulnar crest of the humerus on the proximal margin at about a middle point of the pneumatic fossa.

Relations.—Externally, with the pectoralis major; interiorly, with the pectoralis secundus; internally, with the fascia of the subclavius and the sternum.

Shape.—Fleshy, fusiform.

Action.—Assists the pectoralis secundus in elevating the humerus.

MUSCLES OF THE SCAPULAR REGION

Coraco-humeralis

Scapulo-humeralis

Supraspinatus

Subclavius

Coraco-brachialis

Teres minor

Levator scapulæ

Thoraco-scapularis

Subscapularis

Coraco-humeralis. *Location.*—The scapulo-humeral region.

Origin.—From the outer side of the head of the coracoid, supero-laterally to the long head of the biceps.

Insertion.—To the palmar part of the head of the humerus just inside of the insertion of the pectoralis major.

Shape.—A delicate subcylindrical muscle.

Relations.—Superior to the head of the humerus.

Action.—Aids in extending the humerus.

Scapulo-humeralis (Fig. 14, No. 10). *Location.*—The scapulo-humeral region.

Origin.—From the inner side of the neck of the scapula just within the head of the deltoid.

Insertion.—Passing over the top of the shoulder-joint it is inserted to the palmar part of the humeral head between the insertion of the pectoralis major and the pectoralis secundus.

Shape.—Narrow, flat ribbon.

Relations.—Along the upper margin of the larger portion of the deltoid.

Action.—Aids in extending the humerus.

Supraspinatus (Fig. 19, No. 22). *Location.*—Scapular region.

Origin.—From the superior surface and outer third of the scapula.

Insertion.—To the lower border of the pneumatic fossa of the humerus.

Shape.—Thin, flat, triangular.

Relations.—Externally, with the trapezius and posteriorly, with the teres et infraspinatus.

Action.—If the scapula is the fixed point it will pull the humerus upward and backward.

Subclavius. *Location.*—Anterior sternal region.

Origin.—From the entire outer surface of the sternal process of the sternum and the adjacent outer surfaces of three or four of the hemapophyses.

Insertion.—To the inferior margin of the coracoid bone, the longer fibers passing over to the fossa in the lower third of the posterior part of the coracoid.

Shape.—Rather small, fleshy.

Relations.—Overlapped by the pectoralis tertius.

Action.—Pulls the coracoid outward. Also aids in keeping the coracoid in place.

Coraco-brachialis (Fig. 13, No. 15). *Location.*—Along the coracoid shaft.

Origin.—By a delicate tendon from a small circular point on the postero-mesial part of the shaft of the coracoid immediately above the attachment of the subclavius and the fossa at that point.

Insertion.—To the top of the ulnar tuberosity of the humerus. This subcircular space is common to the teres minor and to this muscle.

Shape.—Long, fusiform.

Relations.—With the shaft of the coracoid between the teres minor and subscapularis.

Action.—To depress the wing.

Teres Minor (Fig. 13, No. 14). *Location.*—The coraco-scapular region.

Origin.—From under the side of the anterior tip of the scapula.

Insertion.—Passing outward behind the coracoid head and beneath the neck of the scapula, it is attached by a small tendon to the top of the ulnar tuberosity of the humerus in common with the coraco-brachialis, with which tendon it fuses.

Shape.—Small, chunky.

Relations.—Anteriorly, with the tip of the scapula and behind the coracoid head. Posteriorly, with the coraco-brachialis.

Action.—Aids in the downward stroke of the wing.

Levator Scapulæ. *Location.*—Scapular region.

Origin.—By two strong digitations, one from each of the lateral processes of the first cervical vertebra.

Insertion.—To the middle third of the inferior part of the blade of the scapula.

Shape.—Small and flat.

Relations.—Inferior to the scapula.

Action.—Pulls the scapula forward, and with it the entire shoulder girdle, which articulates like a hinge-joint at the costo-sternal juncture.

Thoraco-scapularis. *Location.*—Lateral side of the chest.

Origin.—From the outer part of the lower half of the first free rib, the outer side of the next succeeding rib and its epineural appendage, and from a similar surface on the next rib.

Insertion.—By a broad, flat tendon which passes between the two divisions of the subscapularis and inserts at a point at the juncture of the anterior and middle third of the outer margin of the scapula.

Shape.—Broad, flat.

Relations.—Internally, with the ribs and the scapula.

Action.—When the scapula is fixed it is a muscle of inspiration. When the ribs are fixed, it draws the scapula downward.

Subscapularis. *Location.*—Scapulo-humeral region.

Origin.—From the anterior and outer half of the scapula.

Insertion.—By a strong and subcylindrical tendon to the top of the ulnar tuberosity of the humerus close to the combined tendons of the coraco-brachialis and teres minor.

Shape.—A large and powerful muscle.

Relations.—The flat tendon of the thoraco-scapularis divides the posterior portion of the belly into two portions.

Action.—A powerful rotary muscle of the head of the humerus; aids in keeping the humeral head in the glenoid cavity.

THE MUSCLES OF THE BRACHIAL REGION

Biceps

Triceps

Deltoid

Brachialis anticus

Biceps (Fig. 13, No. 8). (Synonym.—Biceps brachii.)

Location.—The anterior brachial region.

Origin.—By a broad tendon it covers the top of the shoulder-joint, dividing into two heads, giving rise to the long and the short heads. The long is inserted into the outer part of the head of the coracoid just beyond the glenoid cavity; the short head to the distal angle of the ulnar tuberosity of the humerus.

Insertion.—To the ulna just in front of the articular cavity for the trochlear surface of the distal extremity of the humerus.

Shape.—Large, subfusiform.

Relations.—On one side by the triceps and the other the deltoid and anteriorly partially covered by the patagii muscles.

Action.—A powerful flexor of the forearm.

Deltoid (Fig. 13, No. 12; Fig. 15, No. 8). *Location.*—Occupying the lateral side of the humerus; the brachial region.

Origin.—Divided into two portions, the long narrow head extends from the clavicular process of the scapula and from the adjacent surface of the same bone. These fibers then extend around the back of the shoulder-joint, and are joined by the fibers that arise from the entire outer surface of the large os humero-scapulare. These latter fibers are inserted upon an extensive area upon the anconal part of the bone beyond the humeral articular head, and to almost the entire shaft below it.

Insertion.—By a subcylindrical tendon to the proximal side of the tubercle of the external condyle, and above the insertion of the extensor metacarpi radialis longus.

Shape.—Large, fleshy, with tendinous attachment.

Relations.—On one side by the biceps and the other by the triceps also the patagii.

Action.—To extend the arm.

Triceps (Fig. 15, No. 7). (Synonym.—Triceps brachii.)

Location.—The posterior humeral region.

Origin.—By three portions: by the internal and external heads, and by the long scapular head. The internal and the external heads are blended except at their proximal extremities. Each head arises from the anconal surface of the shaft of the humerus. The internal head being located toward the deltoid attachment and the external head into the pneumatic fossa. The long head extends from a circumscribed area just posterior to the glenoid cavity of the scapula. The fibers pass around the shoulder-joint, beneath the deltoid.

Insertion.—The broad tendon passes over the elbow-joint and is inserted to the entire under surface of the olecranon of the ulna.

Shape.—Long, large, fleshy.

Relations.—The supraspinatus passes between the internal and the external heads. It is bounded by the biceps and the deltoid.

Action.—An extensor of the antibrachial region directly antagonizing the biceps.

Brachialis Anticus (Fig. 15, No. 9; Fig. 14, No. 5). *Location.*—In the flexure of the humerus and the ulna.

Origin.—It arises fleshy from a circumscribed area on the inner side of the anconal part of the distal extremity of the humerus.

Insertion.—The fibers passing directly over the elbow-joint become inserted to the lateral surface of the proximal end of the ulna close to the margin of the sigmoid cavity.

Shape.—Small, fleshy.

Relations.—With the joint.

Action.—Assists in flexing the forearm upon the arm. Protects the structures in the anterior part of the elbow-joint.

THE MUSCLES OF THE FOREARM AND THE HAND

Extensor metacarpi radialis longior

Supinator brevis

Extensor digitorum communis

Pronator brevis

Flexor metacarpi radialis

Extensor ossis metacarpi pollicis

Pronator longus

Extensor indicis longus

Flexor digitorum sublimis

Flexor digitorum profundus

Flexor carpi ulnaris

Flexor carpi ulnaris brevior

Extensor Metacarpi Radialis Longior (Fig. 13, No. 17). *Location.*—Superior to the radius.

Origin.—By two strong tendinous heads; the outer from the tubercle of the external condyle of the humerus, just above the origin of the tendon of the tensor patagii brevis; the inner and stronger portion from the tubercle found above the oblique trochlear facet of the distal end of the radius.

Insertion.—Becoming a flat, broad, and strong tendon at about the middle of the forearm, finally becomes inserted to the apex of the anchylosed first metacarpal of the carpo-metacarpus.

Shape.—A thick, fusiform belly, tendinous at both extremities.

Relations.—By the side of the pronator brevis.

Action.—Raises the hand, and draws it forward toward the radial margin of the forearm and retains it on the same side.

Extensor Digitorum Communis (Fig. 15, No. 2). *Location.*—The upper metacarpal region.

Origin.—Immediately below the tubercle of the external condyle of the humerus.

Insertion.—By a small tendon to the outer side of the base of the pollex, and at a mid-point upon the anterior rim of the proximal phalanx of the middle finger.

Shape.—Small spindle-shaped, becoming tendinous at about the middle of the forearm.

Relations.—Occupies the middle of the group of muscles on the outer side of the forearm.

Action.—An extensor of the digit.

Supinator Brevis (Fig. 14, No. 7). *Location.*—The superior part of the radius.

Origin.—From the external condyle of the humerus below the origin of the tendon of the extensor digitorum communis.

Insertion.—To the outer side of the shaft of the radius, for nearly one-third of its length.

Shape.—Thin, slender.

Relations.—Inferiorly, with the bone.

Action.—A supinator of the radial region, and antagonistic to the pronators.

Flexor Metacarpi Radialis (Fig. 15, No. 10). *Location.*—The lowest of the group of three muscles on the outer part of the forearm.

Origin.—By two tendinous heads, the longer from the external condyle of the humerus, and the other just beyond the base of the olecranon.

Insertion.—Becomes tendinous at about the middle of the shaft of the ulna and inserts to a prominent process on the proximal third of the posterior part of the shaft of the mid-metacarpal.

Shape.—Fusiform.

Relations.—Inferiorly, the anconeus and superiorly, the extensor digitorum communis.

Action.—A powerful flexor of the hand.

Pronator Brevis (Fig. 13, No. 18). *Location.*—Supero-lateral side of the forearm.

Origin.—From just above the internal condyle of the humerus.

Insertion.—The tendon passing obliquely across the interosseous space and is inserted to the ulnar side of the shaft of the radius, just beyond the juncture of the proximal and the middle thirds.

Shape.—Fusiform.

Relations.—Superior to the pronator longus.

Action.—Pronates the forearm, and flexes the forearm upon the arm.

Pronator Longus (Fig. 14, No. 6). *Location.*—Lateral side of the radio-ulnar region.

Origin.—From the middle of the internal condyle of the humerus.

Insertion.—To the shaft of the radius just beneath the pronator brevis.

Shape.—Massive, ellipsoidal.

Relations.—Between the pronator brevis and flexor digitorum profundus.

Action.—A pronator.

Extensor Ossis Metacarpi Pollicis (Fig. 13, No. 23). *Location.*—Slightly interposed between the ulna and radius.

Origin.—Immediately in front of the greater sigmoid cavity of the ulna.

Insertion.—To the palmar side of the base of the first metacarpal, in common with the extensor metacarpi radialis longior.

Shape.—Delicate, straight.

Relations.—Superiorly, the extensor indicis longus, and inferiorly, the anconeus and the flexor digitorum profundus.

Action.—Extends the hand upon the forearm.

Anconeus (Fig. 15, No. 4). *Location.*—Between the ulna and the radius.

Origin.—By a short, strong, subcylindrical tendon from the lower posterior of the external condyle of the humerus.

Insertion.—To the latero-radial side of the ulna somewhat beyond its middle.

Shape.—Fusiform.

Relations.—Superiorly, the extensor indicis longus, and inferiorly, the ulna.

Action.—A flexor of the forearm.

Extensor Indicis Longus (Fig. 13, No. 22; Fig. 15, No. 13). *Location.*—Infero-lateral to the radius.

Origin.—From about one-half the surface of the proximal portion of the radius.

Insertion.—Possesses a long tendon which passing over a groove at the distal end of the ulna and receiving muscular fibers at the base of the metacarpus, extends down the anterior part of the hand and is inclosed in a sheath in front of the superior part of the first phalanx of the index-finger. This tendon is inserted into the anterior upper rim of the distal phalanx.

Shape.—Small, thin.

Relations.—Supero-laterally with the radius.

Action.—Extensor of the digit.

Flexor Digitorum Sublimis. *Location.*—The forearm.

Origin.—From the internal condyle of the humerus.

Insertion.—Passing over the wrist, it is inserted to the middle phalanx of the hand.

Shape.—A musculo-tendinous band, rather delicate and rudimentary.

Relations.—Closely adhering to the integument that stretches from the internal condyle of the humerus to the wrist.

Action.—Assists these muscles. A flexor.

Flexor Digitorum Profundus (Fig. 13, No. 21). *Location.*—Lateral side of the ulna.

Origin.—By two heads from the proximal extremity of the ulna, between which the brachialis anticus passes.

Insertion.—The two bellies uniting are attached to the under side of the shaft of the ulna. It becomes tendinous about the middle of the shaft, and inserts to the ulnar side of the base of the distal joint of the index-finger.

Shape.—A fleshy belly with long tendon.

Relations.—Inferiorly, with the flexor carpi ulnaris, superiorly, with the pronator brevis.

Action.—A flexor of the digit.

Flexor Carpi Ulnaris (Fig. 13, No. 19). *Location.*—Inferior part of the forearm.

Origin.—By two strong tendons; one from the side and posterior of the internal condyle of the humerus, passing through the humero-ulnar pulley at the side of the base of the olecranon process; the other, to the posterior of the internal condyle of the humerus. It does not pass through a pulley.

Insertion.—A subcylindrical tendon extends back of the ulnare ossicle of the carpus giving off a tendinous slip to the flexor digitorum profundus, and becomes inserted to the anterior rim of the proximal phalanx of the index digit. It has an attachment at the carpus.

Shape.—A strong fleshy belly terminating in a long tendon.

Relations.—The outermost inferior muscle of the forearm.

Action.—A powerful flexor of the hand upon the forearm.

Flexor Carpi Ulnaris Brevior (Fig. 13, No. 20). *Location.*—Inferior part of the antibrachial region.

Origin.—From a broad area on the middle third of the upper side of the shaft of the ulna.

Insertion.—Near the top of the outer edge of the anchylosed os magnum of the carpo-metacarpus.

Shape.—Thin, somewhat flat.

Relations.—To the inside of the flexor digitorum profundus.

Action.—Flexes the hand upon the forearm and rotates the hand toward the body.

OTHER MUSCLES OF THE DIGITS

Extensor proprius pollicis

Flexor minimi digiti

Abductor minimi digiti

Interosseous palmaris

Flexor brevis pollicis

Flexor minimi digiti brevis

Interosseous dorsalis

Flexor metacarpi brevis

Extensor Proprius Pollicis (Fig. 13, No. 29a). *Location.*—Anterior to the radio-carpal joint.

Origin.—From the ulnar side of the tendon of the extensor metacarpi radialis longior.

Insertion.—To the antero-ulnar side of the pollex.

Shape.—Small, spindle-shaped; tendinous at both attachments.

Relations.—Inferiorly, with the proximal and anterior part of the metacarpus.

Action.—An extensor of the pollex.

Flexor Brevis Pollicis (Fig. 13, No. 28a). *Location.*—Anterior to the pollex.

Origin.—From the shaft of the mid-metacarpal bone, just below the pollex.

Insertion.—Distal apex of the pollex.

Shape.—Short, small, fleshy.

Relations.—Superiorly, with the inferior part of the pollex.

Action.—Flexes the pollex.

Flexor Minimi Digiti. *Location.*—Inferior side of the metacarpus.

Origin.—From the posterior part of the median metacarpal close to the os carpi ulnare. A few fibers extend to the ulna.

Insertion.—Into the base of the median phalanx.

Shape.—Small, slender, short.

Relations.—Superiorly, with the metacarpal bone.

Action.—A flexor of the digit.

Flexor Minimi Digiti Brevis (Fig. 13, No. 30). *Location.*—Inferior to the third, or small finger.

Origin.—From the lower and posterior end of the median metacarpal and from the adjacent tendon of the preceding.

Insertion.—Inserted to the apex of the small finger.

Shape.—Small and rather rudimentary.

Relations.—Inferior to the small finger.

Action.—Acts as a posterior ligament of this joint.

Abductor Minimi Digiti. *Location.*—Supero-lateral side of the small finger.

Origin.—From the proximal extremity of the anterior aspect of the third finger.

Insertion.—To the posterior border of the proximal phalanx of the second, or index-finger at a point just above the tip of the third finger.

Shape.—A few muscular fibers with some tendinous material.

Relations.—At the base of the median segment of the index-finger and to the side of the third.

Action.—Antagonizes the powerful flexors on the back of the joint.

Interosseus Dorsalis (Fig. 13, No. 31). *Location.*—By the side of the metacarpus.

Origin.—From the margin of the shaft of the median metacarpal.

Insertion.—To the anterior part of the base of the distal phalanx of the index digit.

Shape.—Short, delicate, with long tendon.

Relations.—Internally, the metacarpal bone.

Action.—Extends the last bone of the middle finger.

Interosseus Palmaris (Fig. 13, No. 32). *Location.*—On the opposite side of the bone from the interosseous dorsalis.

Origin.—From the shaft of the median metacarpal under the preceding.

Insertion.—To the apex of the last joint of the second, or middle finger.

Shape.—Small, delicate, with long slender tendon.

Relations.—Beneath the preceding and along the metacarpal bone.

Action.—Flexes the terminal phalanx.

Flexor Metacarpi Brevis. *Location.*—To the lateral side of the metacarpal bone.

Origin.—From the outer side of the distal extremity of the ulna.

Insertion.—To the base, in front, and on the ulnar side of the proximal phalanx of the index digit.

Shape.—Delicate, with long slender tendon.

Relations.—On the lateral side of the ulna and the metacarpal bone.

Action.—Flexes the metacarpus.

MUSCLES OF THE POSTERIOR LIMB

Sartorius

Gluteus medius

Extensor femoris

Biceps flexor cruris

Semitendinosus accessorius

Femoro-caudal

Obturator internus

Abductor longus

Gluteus primus

Gluteus minimus

Vastus internus

Semitendinosus

Semimembranosus

Obturator externus

Gemellus

Adductor magnus

Sartorius (Fig. 23, No. 7). *Location.*—Extreme anterior portion of the thigh.

Origin.—From the outer two-thirds of the superior surface of the raised emarginations of the anterior border of the ilium, and by fascia from the neural spines of the fourth dorsal vertebra.

Insertion.—Obliquely by a semitendinous fascia to the inner edge of the ligamentum patella, and to the inner and adjacent border of the anterior half of the summit of the tibia.

Shape.—Large, fleshy.

Relations.—Posteriorly, with the gluteus primus and the vasti.

Action.—Extends the leg, flexes and adducts the thigh.

Gluteus Primus (Fig. 23, No. 8 and 9). *Location.*—The outer flat massive muscle of the thigh.

Origin.—By a thin fascia from nearly the entire length of the supero-internal margin of the pre-acetabular portion of the ilium, above the antitrochanter, and from the entire length of the post-acetabular ridge.

Insertion.—Near the patella it joins the tendon of the extensor femoris, and by aponeurosis spreads over the knee and is inserted to the crest of the upper border of the tibia, the patella being incorporated in its aponeurotic ligament.

Shape.—Triangular in shape, it is aponeurotic anteriorly and more fleshy posteriorly.

Relations.—Anteriorly, covers over the posterior portion of the sartorius and the body of the gluteus medius; posteriorly, bordered by the semitendinosus and the biceps.

Action.—Abducts the thigh.

Gluteus Medius (Fig. 24, No. 11). *Location.*—Supero-anterior part of the ilium under the gluteus primus.

Origin.—From the entire supero-internal margin of the pre-acetabular surface of the ilium and the concave surface on the adjacent bone.

Insertion.—By a strong tendon which passes over a bursa on the anterior rim of the trochanter and is inserted obliquely to this trochanter.

Shape.—A fleshy muscle filling the concavity of the pre-acetabular division of the ilium.

Relations.—Superiorly, with the gluteus primus; inferiorly, with the gluteus minimus.

Action.—Abducts and pulls the femur forward.

Gluteus Minimus (Fig. 24, No. 12). (Synonym.—Gluteus profundus.)

Location.—Beneath the medius and the smaller of the two.

Origin.—To the anterior margin of the outer border of the ilium, and from the supero-external surface of the last rib.

Insertion.—The fibers extending downward and backward and outward become tendinous and are inserted below the trochanter of the outer part of the upper third of the femur.

Shape.—Small, fleshy.

Relations.—Inferiorly, with the bone and superiorly, with the gluteus medius.

Action.—Abducts and pulls the thigh forward.

Extensor Femoris (Fig. 24, No. 15). *Location.*—Anterior femoral region. It is divided into two parts: the vastus externus and the crureus.

Origin.—The vastus extemus (Fig. 24, No. 14) arises tendinous from the base of the trochanter of the outer part of the femur and from this point down the bone approximately to the condyle. The crureus (Fig. 24, No. 13) originates by a tendon from the anterior upper prominent rim of the trochanter, from this point extending down the antero-external part of the shaft of the femur.

Insertion.—Merges with the gluteus primus, and, by a fascia-like arrangement, spreads over the front of the knee-joint inserts to the patella and is also inserted into the crest of the tibia.

Shape.—Large, fleshy.

Relations.—Posteriorly, with the femur and anteriorly, with the sartorius.

Action.—Extends strongly the leg upon the thigh.

Vastus Internus (Fig. 16, No. 3). *Location.*—Internal and lateral side of the femur.

Origin.—From the postero-internal part of the shaft of the femur just below the head, and in a straight line extending down the femur.

Insertion.—By a broad tendon along the thickened inner border of the summit of the tibia.

Shape.—Thick, long.

Relations.—Anteriorly, with the bone; surrounded by the other crural muscles.

Action.—Extends the leg upon the knee; a powerful assistant of the extensor femoris.

Biceps Flexor Cruris (Fig. 17, No. D, 1). *Location.*—Posterior tibial region.

Origin.—By a tendinous fascia from the post-acetabular ridge, extending between the antitrochanter and the anterior point of the insertion of the semitendinosus.

Insertion.—To the tuberosity on the external part of the shaft of the fibula a short distance below the head.

Shape.—Large, rather cone-shaped with base directed upward.

Relations.—Anteriorly, with the vastus externus; posteriorly, with the semitendinosus. The inferior tendinous portion passes through a tendinous sling or pulley called the biceps band.

Action.—Flexes the leg upon the thigh.

Semitendinosus (Fig. 23, No. 10). *Location.*—Posterior femoral region.

Origin.—By a tough, strong fascia from the surface of the caudal muscles and from the posterior third of the post-acetabular ridge.

Insertion.—The fibers pass downward and forward, and insert to the tendinous raphe along the posterior margin of the semitendinosus accessorius; lower down it merges with the median fascia of the inner head of the gastrocnemius.

Shape.—Fleshy, broad, flat and long.

Relations.—Internally, with the semimembranosus and anteriorly, with the biceps flexor cruris.

Action.—Flexes the leg.

Semitendinosus Accessorius. *Location.*—Posterior femoral region.

Origin.—From an oblique line just above the condyle, on the posterior of the shaft of the femur.

Insertion.—Its fibers, passing upward and backward, attach themselves to the tendinous raphe common to this muscle and the semitendinosus, and are finally inserted to the inner side of the shaft of the tibia.

Shape.—Flat, oblong.

Relations.—Posteriorly to the shaft of the femur, and externally to the long adductors of the thigh.

Action.—Aids the preceding in flexing the leg.

Semimembranosus (Fig. 24, No. 17). *Location.*—Postero-internal to the semitendinosus.

Origin.—From the outer surface of the ischium, beginning at the lower margin of its notch on the posterior pelvic border, extends on a curved line on the adjacent surface beyond.

Insertion.—The fibers, passing downward and forward, insert by a broad, thin tendon to the shaft of the tibia a short distance below the head and on a line parallel to the long axis of the tibia.

Shape.—Long, narrow, ribbon-shaped.

Relations.—Lies adjacent to the semitendinosus and in the same plane.

Action.—Directly flexes the leg.

Femoro-caudal (Fig. 24, No. 8). *Location.*—Infero-lateral to the caudal and post-femoral region.

Origin.—By a delicate tendon from the base of the pygostyle.

Insertion.—By a thin, flat tendon upon the outer part of the shaft of the femur below the trochanter and at about the juncture of the upper and middle third of the bone.

FIG. 24.—Second layer of muscles of the thigh of a cock. Outside view. 1, Appendico-costales. 2, Levatores costarum. 3, Sacro-lumbalis. 4, Longissimus dorsi. 5, Intercostales. 6, Longissimus dorsi. 7, Latissimus dorsi. 8, Femoro-caudal. 9, Edge of obturator externus. 11, Gluteus medius. 12, Gluteus minimus. 13, Crureus. 14, Vastus externus. 15, Extensor femoris. 16, Biceps flexor cruris. 17, Semimembranosus. 18, Adductor magnus. 19, Adductor longus. 20, External abdominal oblique. 21, Depressor coccygis. 22, Tibialis anticus. 23, Tibialis posticus. 24, Flexor perforans digitorum

profundus. 25, Flexor perforatus medius secundus pedis. 26, Flexor longus hallucis.

Shape.—Long, narrow, spindle-shaped, flattened from side to side.

Relations.—Interiorly to the obturator externus, and along the upper border of the long adductors of the thigh and the semimembranosus.

Action.—Pulls the tail down and to one side.

Obturator Externus (Fig. 24, No. 9). *Location.*—External posterolateral side of the pelvis.

Origin.—From the posterior half of the periphery of the ischiatic foramen and the concavity found on the external surface of the lateral part of the pelvis.

Insertion.—The fibers pass across to the femur and are inserted by a broad, flat tendon to the shaft of the femur just below the trochanter.

Shape.—Thick, fleshy.

Relations.—Lies just above the femoro-caudal muscle, the sciatic nerve and femoral artery pass over and external to its tendon.

Action.—Pulls the head of the femur backward.

Obturator Internus (Fig. 16, No. E, 1). *Location.*—Occupies the space between the ischium and the pubis.

Origin.—From the ventral surface of the ischium, including the posterior border, from the inner line of the corresponding post-pubis, and from the membrane covering the space between these two bones.

Insertion.—Becoming dense, subcylindrical, and strong, it passes through the obturator foramen and is inserted to the outer part of the trochanter of the femur, in common with the gemellus and opposite to the insertion of the gluteus medius.

Shape.—Subtriangular and flat with a central tendon; bipenniform.

Relations.—Outside the pelvis the tendon lies upon the gemellus and is external to it. Fills the obturator foramen and the space between the ischium and the pubis.

Action.—Acts as a posterior stay to the head of the femur.

Gemellus (Fig. 25, No. 5). *Location.*—Posterior to the head of the femur.

Origin.—From the fossa between the acetabulum and the obturator foramen, and on the outer side of the pelvis.

Insertion.—Its fibers are attached by fascia to the tendon of the obturator internus, and pass directly to the trochanter of the femur where they insert in common with that of the obturator internus.

Shape.—Strong, thick, chunky.

Relations.—Postero-external to the head of the femur.

Action.—Like the two preceding, when the head of the femur is fixed, it pulls the pelvis forward and steadies it on the head of the femur.

Adductor Longus (Fig. 24, No. 19). *Location.*—In the posterior femoral region.

Origin.—From a line on the lateral part of the pelvis.

Insertion.—The fibers, passing downward and forward, are inserted on a longitudinal line along the posterior part of the shaft of the femur.

Shape.—Broad, flat, fleshy.

Relations.—Posteriorly to the femur, and anteriorly to the adductor magnus.

Action.—Adducts the thigh.

Adductor Magnus (Fig. 24, No. 18). *Location.*—Posterior femoral region.

Origin.—From the fine constituting the lower boundary of the ischiatic fossa on the outer lateral part of the pelvis.

Insertion.—The fibers pass downward and are inserted to the superior curve of the internal femoral condyle. The internal head of the gastrocnemius blends with this muscle just above its attachment.

Shape.—Long, narrow.

Relations.—Posteriorly to the adductor magnus and anteriorly to the semimembranosus.

Action.—Strongly adducts the thigh.

THE TIBIAL GROUP

Gastrocnemius

Peroneus longus

Extensor longus digitorum

Tibialis posticus

Flexor longus hallucis

Flexor perforatus annularis primus pedis

Flexor perforatus medius secundus pedis

Flexor perforans digitorum profundus

Soleus

Tibialis anticus

Extensor hallucis brevis

Flexor perforatus indicis secundus pedis

Flexor perforatus medius primus pedis

Flexor perforatus indicis primus pedis

Extensor brevis digitorum

Flexor hallucis brevis

Extensor annularis brevis

Gastrocnemius (Fig. 23, No. 11). *Location.*—Posterior tibial region.

Origin.—There are three heads: the internal, the external, and the tibial. The external head extends by a short, flattened, strong tendon from the postero-external part of the external condyle of the femur; the internal head extends from the outer surface of the inner condyle of the femur; and the tibial head extends from the entire inner rim of the tibial summit and from the free edge of the adjacent crest.

Insertion.—At the lower fourth of the tibial shaft, terminates in a broad, flat tendon which passes over the shallow, longitudinal groove of the tibial cartilage, and, crossing the tibio-tarsal joint, is inserted to the posterior surface of the hypotarsus of the metatarsal bone, and finally, below this point, merges into the podothecal sheath confining the flexor tendons.

Shape.—Large, fleshy, somewhat cone-shaped with the base upward.

Relations.—The posterior fleshy muscle of the post-tibial region located anterior to the soleus.

Action.—Extends the metatarsus on the tibia.

Soleus (Fig. 25, No. G, 1). *Location.*—The posterior tibial region.

Origin.—From the posterior part of the head of the tibia.

Insertion.—By a long slender tendon to the proximal end and toward the inner angle of the tibial cartilage, some of the fibers passing to the tendon of the gastrocnemius.

Shape.—Small, flattened.

Relations.—Anteriorly, with the flexor perforans digitorum profundus and posteriorly, with the gastrocnemius.

Action.—Similar to that of the gastrocnemius.

Peroneus Longus (Fig. 23, No. 12). *Location.*—Antero-lateral tibial region.

Origin.—From the raised crest in front of the head of the tibia, and from the fascia that covers the outer side of the knee-joint.

Insertion.—The fibers, passing downward, then downward and outward, terminate in a small tendon at the lower third of the tibia. This tendon bifurcates just above the tibial condyles at the outer part of the limb, the shorter and stronger attaches to the fibrous fascia covering the tibial cartilage, the other merging with the tendon of the flexor perforatus medius primus pedis about 1 centimeter below the hypotarsus of the tarso-metatarsus.

Relations.—Covers over all the muscles of the anterior tibial region.

Shape.—A broad muscular sheet.

Action.—Assists the flexor perforatus medius primus pedis.

Tibialis Anticus (Fig. 24, No. 22). *Location.*—In the anterior tibial region.

Origin.—The muscular belly may be easily divided into two parts. The inner head extends from the head of the tibia immediately beneath the peroneus longus. The outer head arises by a strong tendon from a depression on the antero-inferior ridge of the outer condyle of the femur.

Insertion.—The tendon passing through the fibrous ligamentous loop just above the tibial condyles, inserts to a tubercle on the shaft just below the head of the tarso-metatarsus.

Shape.—Large, fusiform.

Relations.—Beneath the peroneus longus.

Action.—Flexes the metatarsus upon the tibia.

Extensor Longus Digitorum (Fig. 25, No. G, 6). *Location.*—Anterior tibial region.

Origin.—From the inferior portion of the crest and a portion of the shaft of the fibula.

Insertion.—It becomes tendinous at the lower third of the bone, passing under the bony ridge just above the condyle in front, and over the ankle-joint. It is bound down by firm fascia, and at the trochlea of the basal toe joints, divides into three small tendinous slips. These pass over the superior part of the second, the third, and the fourth toes, bifurcating in their course, and are inserted to the distal ones.

Shape.—Long, penniform, with a long, slender tendon.

Relations.—Immediately below the tibialis anticus.

Action.—Extends the digits as their long extensor.

Extensor Hallucis Brevis (Fig. 25, No. G, 4). *Location.*—The anterior metatarsal region, along the antero-internal edge.

Origin.—From just below the summit of the antero-internal part of the head of the tarso-metatarsus, and from the adjacent shaft below, and from the tendon of the tibialis anticus.

Insertion.—Into the process at the superior part of the bone of the bony claw of the hallux.

Shape.—Slender, thin in diameter, with a long, slender tendon.

Relations.—Superiorly, with the antero-internal edge of the metatarsus and interiorly, its tendon with the top of the hallux.

Action.—Extends the hallux.

Extensor Brevis Digitorum. *Location.*—On the anterior surface of the metatarsus.

FIG. 25.—Inner layer of muscles of the thigh of a hen. Outside view.

A. Pad of fibrocartilage or meniscus of femoro-tibial articulation. 1, Inner semi-lunar fibrocartilage. 2, External semi-lunar cartilage. 3, Anterior border.

B. Tarsal meniscus or pad of fibrocartilage.

C. Articular surface of the tarsal joint. 1, Posterior ligamentum cruciatum genu. 2, Femoral tendinous insertion of tibialis anticus. 3, Ligamentum tibio-fibulare. 4, Ligamentum interosseum. 5, Ligamentum transversum of the meniscus. 6, Anterior ligamentum cruciatum genu.

D. 1, Tibia. 2, Fibula. 3, Femur. 4, External ligamentum laterale genu.

E. Inferior surface of the tail of a hen. 1, Infracoccygis. 2, Lateralis coccygis.

F. Outside pelvic view. 1, Obturator ligament. 2, Tendon of the obturator internus. 3, Ligamentum oblongum. 4, Ligamentum capsulare. 5, Gemellus.

G. Outside view of leg of a cock. 1, Soleus. 2, Flexor longus hallucis. 3, Flexor perforans digitorum profundus. 4, Extensor brevis hallucis. 5, Extensor brevis annularis. 6, Extensor longus digitorum. 7, Tendon of the tibialis posticus.

H. 1, Transversalis abdominalis. 2, Cotyloid cavity.

Origin.—By a small fleshy belly to the anterior part of the proximal end of the metatarsus.

Insertion.—It becomes tendinous about the middle third of the metatarsus. Extending down the anterior face of the metatarsus it is inserted to the inner tubercle of the base of the first phalanx of the outer or fourth toe.

Shape.—Long, slender.

Relations.—Posteriorly with the metatarsus.

Action.—Pulls the fourth toe upward and inward.

Flexor Hallucis Brevis. *Location.*—Postero-internal side of the metatarsus.

Origin.—Side and lower margin of the inner aspect of the hypotarsus of the tarso-metatarsus and from the shaft of this bone immediately below it.

Insertion.—Winds round inner side of basal joint of the hallux and is inserted on its under side.

Shape.—Small flat muscle gradually tapering into a tendon.

Relations.—Posteriorly with the metatarsus.

Action.—To flex the hallux.

Extensor Annularis Brevis. *Location.*—On postero-external side of metatarsus.

Origin.—External aspect of the hypotarsus and the shaft below.

Insertion.—Supero-external part of the basal phalanx of the fourth toe.

Shape.—Small flat muscle.

Relations.—Posteriorly with the metatarsus.

Action.—Extensor of the fourth toe.

Tibialis Posticus (Fig. 24, No. 23). *Location.*—In the posterior tibial region.

Origin.—From the whole shaft of the fibula below the insertion of the biceps flexor cruris, and from the shaft of the tibia a short distance below this insertion. Also from the interosseous membrane, and the adjacent surface of the tibial shaft.

Insertion.—Passing in front of the external malleolus, crossing the ankle-joint, finally inserted into the outer edge of the summit of the tarso-metatarsal bone.

Shape.—Long, subcylindrical.

Relations.—Medially, with the flexor perforans digitorum pedis; superiorly, with the postero-external portion of the tibia.

Action.—Extends the metatarsus upon the tibia.

Flexor Perforatus Indicis Secundus Pedis (Fig. 17, No. 5). *Location.*—The posterior tibial region.

Origin.—From the external surface of the outer condyle of the femur, just below the head of the gastrocnemius.

Insertion.—Passing in an oblique manner along posterior to the tibia, extends through the superficial part of the tibial cartilage to the outer side of the leg; then crossing the tibio-tarsal joint, it passes through a special canal of the hypotarsus, down the back of the tarso-metatarsus, under the annular ligament of the sole of the foot and a cartilaginous structure below this ligament, over the trochlea, and is inserted to the middle shaft of the second phalanx of the second toe.

Shape.—Thin, flat, broadly spindle-shaped.

Relations.—Posteriorly, with the flexor perforatus medius secundus pedis, and anteriorly, with the flexor perforatus annularis primus pedis.

Action.—Flexes the toes.

Flexor Longus Hallucis (Fig. 17, No. 7; Fig. 25, No. 2). *Location.*—The posterior tibial region.

Origin.—By two tendinous heads: one from the posterior part of the femur, just below the two condyles; the other from the outer part of the external condyle of the femur. The deep flexor passes between these two heads.

Insertion.—Becoming tendinous at the lower third of the leg, passes beneath the superficial flexors in a special canal on the outer side of the tibial cartilage, and extends through the large canal of the hypotarsus (Fig. 12, No. *F*, 5) next to the shaft, then down to the apex of the accessory metacarpal bone. It is inserted to the tubercle on the inferior proximal end of the ungual phalanx.

Shape.—Large, fusiform.

Relations.—With the exterior of the deep flexors.

Action.—Flexes the hallux as its long flexor.

Flexor Perforatus Annularis Primus Pedis (Fig. 17, Nos. *D*, 8 and *C*, 6). *Location.*—The posterior tibial region.

Origin.—From the inner side of the belly of the flexor longus hallucis.

Insertion.—Becoming tendinous at the lower third of the tibia, passes a little to the outer side and through the tibial cartilage, then over the ankle-joint and through the hypotarsal canal (Fig. 12, No. 5) to the under side of the outer toe, where it forms a sheath through which the deep flexors extend. It gives off on either side a tendinous slip which attaches to the basal phalanx.

Shape.—Long, slender, compressed laterally.

Relations.—Internally, with the flexor perforatus indicis secundus pedis, and externally, with the flexor longus hallucis.

Action.—Flexes the toes as their long flexor. A second flexor of the outer toe.

Flexor Perforatus Medius Secundus Pedis (Fig. 17, No. 9). *Location.*—The posterior tibial region.

Origin.—From a strong fascia that merges with the knee-joint, and by a tendon from the external condyle of the femur in common with the tendon of the flexor perforatus indicis secundus pedis.

Insertion.—Continued as a flattened tendon extending from the lower third of the leg, passing obliquely through the tibial cartilage and then through the interno-posterior canal of the hypotarsus, (Fig. 12, No. F, 5) inserted to the second joint of the middle toe.

Shape.—Long, fusiform.

Relations.—With the shaft of the fibula, which it covers.

Action.—Flexes the middle toe.

Flexor Perforatus Medius Primus Pedis (Fig. 17, No. 2). *Location.*—Posterior tibial region.

Origin.—From the inner side of the muscular portion of the flexor perforatus annularis primus pedis.

Insertion.—Terminating in a flattened tendon, passes through the tibial cartilage and hypotarsus, and then extends along the shaft of the basal joint of the third toe, and, dividing into two slips, inserts to either side of its shaft.

Shape.—Long, rather small.

Relations.—With the flexor perforatus annularis primus pedis on the inner side.

Action.—Flexes the third toe.

Flexor Perforatus Indicis Primus Pedis (Fig. 17, No. 4). *Location.*—The posterior tibial region.

Origin.—From a thin, broad tendon, on the external condyle of the femur, in common with the flexor perforatus medius secundus pedis.

Insertion.—Becoming tendinous, passes through the tibial cartilage and the hypotarsus of the metatarsus (Fig. 16, *A*) and inserts to the sides of the basal joint of the second toe.

Shape.—Long, slender, fusiform; somewhat flattened laterally.

Relations.—With the flexor perforatus medius secundis pedis, posteriorly.

Action.—A flexor of the toes.

Flexor Perforans Digitorum Profundus (Fig. 24, No. 24 and Fig. 25, No. G, 3). *Location.*—The posterior tibial region.

Origin.—From the posterior part of the external condyle of the femur, from the posterior part of the tibia just below the summit, and from a point near the origin of the soleus.

Insertion.—Its heavy flattened tendon passes through the tibial cartilage, this tendon dividing, the branches pass along the under side of the toes, perforating the outer tendons and become inserted to the proximal tubercle at the base of the under side of the ungual phalanxes of the second, third, and fourth toes.

Shape.—Long, fleshy, with a long tendon.

Relations.—Closely with the tibial bone; the deepest of all the flexors.

Action.—Flexes the digits.

THE MUSCLES OF THE EYE

Orbicularis palpebrarum

Depressor palpebræ inferioris

Pyramidalis nictitans

Obliquus inferior

Rectus inferior

Levator palpebræ superioris

Quadratus nictitans

Obliquus superior

Rectus superior

Rectus externus

Rectus internus

Orbicularis Palpebrarum (Fig. 7, No. 30). (Synonym.—Orbicularis oculi.)

Location.—Upon the lacrimal and maxillary bones.

Origin.—From the lacrimal and maxillary bones and the ciliary margin.

Insertion.—To the lower free edge of the tarsal cartilage.

Shape.—Thin, delicate layer of fibers.

Relations.—Externally, with the skin; internally, with the lacrimal and the maxillary bones.

Action.—Aids in closing the eye.

Levator Palpebræ Superioris (Fig. 7, No. 24). *Location.*—Along the superior roof of the orbit.

Origin.—Along a longitudinal line found near the middle of the roof of the orbit.

Insertion.—To the ciliary margin of the upper lid, near the outer canthus.

Shape.—Thin, delicate layer.

Relations.—Superiorly, with the bony wall, and inferiorly, with the eyeball.

Action.—Raises the superior eyelid.

Depressor Palpebræ Inferioris (Fig. 7, No. 29). (Synonym.—Malaris.)

Location.—Along the inferior border of the eye.

Origin.—From the inferior border of the interorbital foramen.

Insertion.—To the lower lid.

Shape.—Thin, flat, delicate.

Relations.—Superiorly, with the eyeball; inferiorly, with the inferior orbital wall.

Action.—Pulls the lower lid down.

Quadratus Nictitans (Fig. 26, No. B, 7). *Location.*—Above the eyeball along the inferior border of the upper wall of the orbit.

Origin.—From the sclerotic from the upper part of the ball.

Insertion.—To the upper part of the sheath of the optic nerve near the optic foramen.

Shape.—Broad, thin, quadrate.

Relations.—Inferiorly, with the eyeball; superiorly, with the orbital wall, with the superior oblique, and with the superior rectus.

Action.—Aids in pulling the nictitating membrane obliquely outward and downward over the forepart of the eyeball.

Pyramidalis Nictitans (Fig. 26, No. B, 8). *Location.*—Inferoposteriorly to the eyeball.

Origin.—From the lower nasal side of the eyeball.

Insertion.—The fibers converge toward the upper part of the optic nerve, into a tendon which passes through a pulley at the free margin of the quadratus. Inserts to the lower part of the margin of the third eyelid, the nictitans.

Shape.—Pyramidal; of thin layers of fibers.

Relations.—Internally, with the bony orbital wall; externally, with the eyeball.

Action.—Aids the quadratus nictitans.

Obliquus Superior (Fig. 26, No. B, 5). (Synonym.—Obliquus dorsalis.)

Location.—Superior to the eyeball.

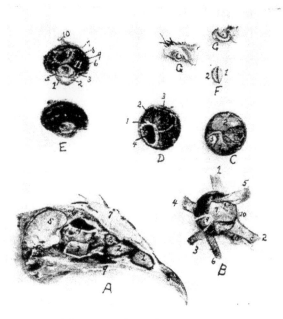

FIG. 26.—Structures of the eye and nasal passage.

A. 1, The anterior turbinated bone. 2, The posterior turbinated bones. 3, Orbital gland. 4, Ophthalmic division of the fifth pair of cranial nerves. 5, Section through the cerebrum. 6, Section through the skull showing the air spaces. 7, Section through the comb. 8, Anterior portion of the upper beak. 9, Edge of the hard palate.

B. The intrinsic muscles of the eyeball. 1, Superior rectus. 2, Internal rectus. 3, Inferior rectus. 4, External rectus. 5, Superior oblique. 6, Inferior oblique. 7, Quadratus. 8, Pyramidalis. 9, Optic nerve. 10, Edge of membrana nictitans.

C. Posterior part of the eye showing at the inferior portion of the ball the pyramidalis muscle at 1, and at 2, the quadratus. 3, Optic nerve.

D. Inner view of the posterior part of the vitreous chamber. 1, Sclerotic coat. 2, Choroid coat. 3, Retina. 4, Pecten.

E. Longitudinal section through the eye of a hen. 1, Cornea. 2, Anterior chamber. 3, Scleral ring. 4, Crystalline lense. 5, Iris. 6, Retina. 7, Sclerotic. 8, Choroid. 9, Pecten. 10, Optic nerve. 11, Vitreous chamber.

F. An edge view of the crystalline lense of a hen showing it to be asymmetrical biconvex. 1, The anterior side. 2, The posterior side.

G. The upper section through the anterior portion of the anterior turbinated bone showing one complete circle and the lower a section through the middle turbinated showing one and one-half turns.

Origin.—From the orbital septum just back of the inner canthus, the fibers passing backward and forward.

Insertion.—Into the sclera, anterior to the sclerotic plates.

Shape.—Small, thin, fan-shaped.

Relations.—With the eyeball, superiorly; at its extremities with the superior rectus, interiorly.

Action.—Rotates the eyeball.

Obliquus Inferior (Fig. 26, No. B, 6). (Synonym.—Obliquus ventralis.)

Location.—Inferior to the eyeball.

Origin.—From the interorbital septum near the inner canthus of the eye.

Insertion.—By a broad expansion to the inferior portion of the eyeball.

Shape.—Tendinous at its insertions; fan-shaped as it passes downward and outward.

Relations.—Internally, with the eyeball; and externally, with the bony orbital wall, with the inferior rectus muscle, inferiorly.

Action.—Aids in rotating the eye; opposes the obliquus superior.

Rectus Superior (Fig. 26, No. B, 1). (Synonym.—Rectus dorsalis.)

Location.—Superior to the eyeball.

Origin.—Along the supero-posterior border of the optic foramen.

Insertion.—To the sclera just within the margin of the bony circle of the eye.

Shape.—Flat, thin.

Relations.—Internally, with the obliquus superior; superiorly, with the bony wall of the orbit.

Action.—To pull the eyeball upward.

Rectus Inferior (Fig. 26, No. B, 3). (Synonym.—Rectus ventralis.)

Location.—Inferior to the eyeball.

Origin.—From the inferior border of the optic foramen.

Insertion.—To the lower portion of the eyeball within the margin of the bony circle of the eye.

Shape.—Thin, fan-shaped.

Relations.—Superiorly, with the eyeball; inferiorly, with the bony wall of the orbit.

Action.—Pulls the eyeball downward.

Rectus Externus (Fig. 26, No. *B*, 4). (Synonym.—Rectus lateralis.)

Location.—External to the eyeball.

Origin.—By two heads infero-laterally to the optic foramen.

Insertion.—To the postero-external side of the eyeball.

Shape.—Short, fan-shaped.

Relations.—Internally, with the eyeball; externally, with the bony wall of the orbit.

Action.—Pulls the eyeball outward.

Rectus Internus (Fig. 26, No. *B*, 2). (Synonym.—Rectus medialis.)

Location.—Inner side of the eyeball.

Origin.—From the supero-anterior border of the optic foramen.

Insertion.—To the sclera on the anterior portion of the eyeball.

Shape.—Thin, fan-shaped.

Relations.—Internally, with the eyeball and with the pyramidalis; externally, with the bony wall of the orbit.

Action.—Pulls the eyeball inward.

THE MUSCLES OF THE EAR

Circumconcha

Tensor tympani

Circumconcha (Fig. 7, No. 32). *Location.*—Surrounds the external ear.

Origin and Insertion.—The circumconcha, surrounding the periphery of the ear, is adherent to the skull and loosely to the skin, and is attached to the outer terminus of the supra-occipital crest.

Relations.—Internally, with the skull; externally, with the skin.

Action.—Relaxes the tympanum.

Tensor Tympani (Fig. 7, No. 31). *Location.*—External to the quadrate and to the external auditory meatus.

Origin and Insertion.—From the surface of the quadrate and the inner end of the quadrato-jugal to the inner surface of the tympanum.

Shape.—A few fibers.

Relations.—Internally, with the quadrate and the quadrato-jugal; externally, the integumental duplicature.

Action.—Tenses the tympanum.

FUNCTIONS OF MUSCLES

To cause the feathers on the top of the head to lie flat:

Dermo-frontalis.

To raise the feathers along the superior part of the neck and along the dorsal region:

Dermo-dorsalis.

To tense the patagial region:

Dermo-tensor patagii,

Dermo-temporalis.

To cause the feathers to lie close to the body:

Dermo-humeralis,

Dermo-pectoralis.

To tense the lateral cervical integument:

Dermo-temporalis.

To support the lingual apparatus and superior larynx:

Platysma myoides.

To manipulate the feathers and skin of the inferior part of the neck:

Cleido-trachealis.

To control the skin in the shoulder region:

Dermo-spinalis.

To cause the feathers of the back to lie close to the skin:

Dermo-iliacus.

To depress the humeral region:

Dermo-ulnaris.

To close the jaw:

Temporalis,

Pterygoideus internus,

Pterygoideus externus,

Masseter.

To open the jaw:

Biventer maxilla,

Entotympanic.

To raise the trachea and hyoid apparatus:

Digastricus.

To elevate the tongue:

Mylo-hyoideus.

To pull the tongue to one side:

Singly[5]—

Cerato-hyoideus,

Stylo-hyoideus.

Alone when the larynx is fixed—

Sterno-hyoideus.

Those which protrude the tongue from the mouth:

Together—

Cerato-hyoideus,

Genio-hyoideus.

To aid in deglutition:

When the base of the tongue is fixed—

Sterno-hyoideus.

To depress the tongue:

Together—

Sterno-hyoideus.

To depress the tip of the tongue and elevate the base:

Depressor glottis.

To elevate the hyoid arches:

Cerato-glossus.

To extend the head:

Complexus,

Rectus capitis anticus minor.

To flex the head upon the neck:

Flexor capitis inferior.

To extend the head on the neck and elevate the neck:

Biventer cervicis.

To raise the neck upward:

Longus colli posticus.

To flex each vertebra on the preceding or succeeding laterally:

Obliquus colli,

Intertransversales.

To pull the neck downward:

Longus colli anterior.

To pull the head downward:

Together—

Rectus capitis lateralis,

Trachelo-mastoideus.

To pull the head downward and to one side:

Singly—

Rectus capitis lateralis,

Trachelo-mastoideus.

To approximate the spinous processes of the vertebræ:

Interspinales.

To approximate the vertebræ in a supero-lateral direction:

Interarticulares.

To flex the vertebræ supero-laterally:

Obliquo-transversales.

To extend the neck:

Together when the first rib is fixed—

Scalenus medius.

To close the glottis:

Constrictor glottidis.

To open the glottis:

Thyreo-arytenoideus.

To approximate the tracheal rings:

Tracheo-lateralis.

To tense the inferior larynx:

Broncho-trachealis posticus,

Broncho-trachealis anticus,

Broncho-trachealis brevis,

Bronchialis posticus,

Bronchialis anticus.

To relax the inferior larynx:

Sterno-trachealis.

To aid in respiration:

When the first rib is fixed—

Intercostales.

When wing is fixed—

Latissimus dorsi,

Dermo-cleido dorsalis.

When the scapula is fixed—

Serratus magnus anticus,

Serratus parvus anticus,

Thoraco-scapularis,

Levatores costarum,

Appendico-costales,

Triangularis sterni.

To flex the shoulder-joint:

Latissimus dorsi.

To elevate the humerus:

Latissimus dorsi.

To draw the scapula forward:

Trapezius,

Rhomboideus.

To close the wing:

Teres et infraspinatus.

To extend the humerus:

Coraco-humeralis,

Scapulo-humeralis.

To pull the humerus upward and backward:

When the scapula is fixed—

Supraspinatus.

To pull the coracoid outward:

Subclavius.

To depress the wing:

Coraco-brachialis.

To depress the scapula:

When the ribs are fixed—

Thoraco-scapularis.

To rotate the humerus:

Subscapularis.

To flex the forearm:

Pronator brevis,

Biceps,

Brachialis anticus,

Anconeus,

Tensor patagii longus,

Tensor patagii brevis.

To extend the arm:

Deltoid.

To extend the antibrachial region:

Triceps.

To raise and to draw the hand forward:

Extensor metacarpi radialis longior.

To extend the digit:

Extensor indicis longus,

Extensor digitorum communis,

Extensor proprius pollicis,

Interosseous dorsalis.

To supinate the radial region:

Supinator brevis.

To flex the hand:

Flexor metacarpi radialis.

To pronate the forearm:

Pronator brevis,

Pronator longus.

To extend the hand on the forearm:

Extensor ossis metacarpi pollicis.

To flex the digit:

Flexor minimi digiti,

Flexor digitorum sublimis,

Flexor brevis pollicis,

Flexor digitorum profundus,

Interosseous palmaris.

To flex the hand upon the forearm:

Flexor carpi ulnaris,

Flexor carpi ulnaris brevior.

To rotate the hand toward the body:

Flexor carpi ulnaris brevior.

To abduct the digits:

Abductor minimi digiti.

To flex the metacarpus:

Flexor metacarpi brevis.

To extend the leg:

Sartorius.

To adduct the thigh:

Sartorius,

Adductor longus,

Adductor magnus.

To abduct the thigh:

Gluteus primus,

Gluteus medius,

Gluteus minimus.

To pull the thigh forward:

Gluteus minimus,

Gluteus medius.

To extend the leg upon the thigh:

Extensor femoris,

Vastus internus.

To extend the leg upon the knee:

Vastus internus.

To flex the leg upon the thigh:

Biceps flexor cruris,

Semitendinosus,

Semitendinosus accessorius,

Semimembranosus.

To pull the tail down and to one side:

Femoro-caudal.

To pull the head of the femur backward:

Obturator externus.

To act as a posterior stay to the head of the femur:

Obturator internus.

To pull the pelvis forward and to steady it on the head of the femur:

Gemellus,

Obturator internus,

Obturator externus.

To extend the metatarsus on the tibia:

Gastrocnemius,

Soleus,

Tibialis posticus.

To flex the toes:

Peroneus longus,

Flexor perforatus medius primus pedis,

Flexor perforatus indicis secundus pedis,

Flexor perforatus medius secundus pedis,

Flexor perforatus indicis primus pedis,

Flexor perforans digitorum profundus,

Flexor longus hallucis,

Flexor perforatus annularis primus pedis.

To flex the metatarsus on the tibia:

Tibialis anticus.

To extend the toes:

Extensor longus digitorum,

Extensor hallucis brevis.

To flex the back laterally and to aid in raising the body.

Sacro-lumbalis,

Longissimus dorsi.

To elevate the tail:

Levator coccygis,

Levator caudæ.

To depress the tail:

Depressor caudæ,

Infracoccygis.

Together—

Depressor coccygis,

Lateralis caudæ.

To pull the tail downward and to one side:

Singly—

Lateralis coccygis.

Depressor coccygis,

Lateralis caudæ.

To close the eyelids:

Orbicularis palpebrarum.

To raise the superior eyelid:

Levator palpebræ superioris.

To pull the lower eyelid downward:

Depressor palpebræ inferioris.

To pull the membrana nictitans over the eyeball:

Quadratus nictitans,

Pyramidalis nictitans.

To rotate the eyeball:

Obliquus superior,

Obliquus inferior.

To pull the eyeball upward:

Rectus superior.

To pull the eyeball downward:

Rectus inferior.

To pull the eyeball outward:

Rectus externus.

To pull the eyeball inward:

Rectus internus.

To tense the ear drum:

Tensor tympani.

To relax the ear drum:

Circumconcha.

To depress the humerus:

Pectoralis major.

To raise the humerus:

Pectoralis secundus,

Pectoralis tertius.

5. Muscles are arranged in pairs. Singly means one muscle acting alone, *i.e.*, without its fellow; together means acting both at the same time.

SPLANCHNOLOGY

The Digestive Apparatus.—This apparatus, apparatus digestorius, consists of the organs directly concerned in the reception of food, in its passage through the body, and in the expulsion of the unabsorbed portion. For convenience, these organs are grouped as follows: the alimentary canal and the accessory organs.

The alimentary canal is a tube which extends from the mouth to the anus. It has a complete lining of mucous membrane, external to which is an almost continuous muscular coat. The abdominal portion of the tube is largely covered with a serous membrane, the visceral peritoneum. The canal consists of the following consecutive segments: mouth, pharynx, first portion of the esophagus, crop, second portion of the esophagus, proventriculus, gizzard, small intestine, large intestine, and cloaca.

The accessory organs are beak, tongue, salivary glands, liver, and pancreas.

The Mouth (Fig. 27).—The distinctive character of the mouth of the fowl consists, in the absence of lips and teeth and instead of jaw bones of other animals, of a *beak*. The edge of the beak is covered by a horn-like gum. The shape of the beak differs in the various classes of birds. In the chicken the beak is short, strong, thick, and pointed; in palmipeds it is flattened. The upper mandible extends out over the lower mandible. The outer portion of the upper and the lower beaks is covered by a horny sheath. There is no velum, and the mouth cavity is extended rather continuously into the pharynx (Fig. 27, No. 4). The posterior cross bar of the hard palate (Fig. 27, No. 5), which possesses a row of filiform papillæ projecting backward, marks the upper boundary line of the mouth; while the posterior end of the tongue (Fig. 27, No. 6), likewise with a row of filiform papillæ, marks its lower boundary line.

Parts of the Mouth.—The mouth of fowls is divided into an upper and a lower half. The upper half is divided into an anterior and a posterior part.

The anterior part extends from the anterior tip of the beak to the posterior nares. It is further divided into an outer hard part and an inner soft part. The outer part is covered by epithelium which may be considered as a continuation from the upper outer portion of the beak. A portion of this is cornified. The inner part is covered by a mucous membrane containing mucous glands, nerves, and blood-vessels.

In the posterior part of the upper half of the mouth, are situated the posterior nares (Fig. 27, No. 10) which appear as a longitudinal slit in the center of the hard palate. There is also found in this part a furrow which contains the opening of the Eustachian tubes, or tubæ auditivæ. There are also two ridges near the same posterior part in which are red and white papillæ, which may appear in rows, and which contain mucous glands.

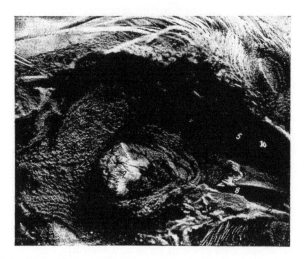

FIG. 27.—Mouth and pharynx of a cock laid open. 1, Eye. 2, Edge of beak covered with a horn-like gum. 3, Tongue. 4, Pharynx. 5, Posterior cross bar of hard palate. 6, Posterior end of tongue. 7, Anterior end of esophagus. 8, Phrenum. 9, Palatal papillaries. 10, Posterior nares.

The lower half of the mouth lies between the lower jaw bones and the walls of the mouth cavity. The tongue is attached here by means of the phrenum (Fig. 27, No. 8). This half of the mouth forms a pocket-like structure which aids in taking up the food.

The mouth cavity is lined with stratified squamous epithelium, continuous with that of the pharynx and the esophagus. Taste buds are located in the mucous membrane of the mouth.

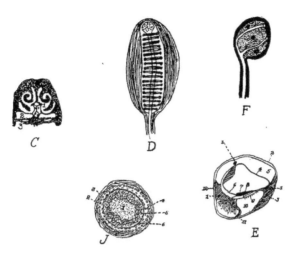

FIG. 28.—Various anatomical parts of the fowl.

C. A section through the nasal region of the fowl. 1, Nasal passages showing the turbinated bones dividing the nasal passage into the superior, middle and inferior meati. 2, The infraorbital sinus. 3, The hard palate.

D. Herbst's touch corpuscle from the beak of a quail.

E. A vertical section through the ductus cochlearis of a pigeon. 1, A bloodvessel. 2, The periosteum. 3, The bony structure. 4, The vascular integument. 5. The scala vestibuli. 6, The inner hyaline cylindrical cells. 7, The membrana tectoria. 8, The papilla acustica basilaris. 9, The membrana basilaris. 10, The scala tympani. 11, The ganglion of the cochlear nerve in the ramus basilaris. 12, The periosteum (Gadow).

F. A corpuscle of the soft papilla of a duck's tongue.

J. A transverse section through a feather papilla. 1, The pulp. 2, The malpighian layer. 3, The corium. 4, The stratum corneum of the papilla. 5, The malpighian cell group of the main shaft. 6, The horny sheath.

In many water fowls, as geese and ducks, the gum edge of the mandible has grooves extending crosswise, in which are numerous terminals of the trigeminus nerve arranged as taste organs (Fig. 29, G and Fig. 28, D and F). In many birds of prey and in water birds, *e.g.*, the goose and the duck, there is found at the base of the beak, a very thin, nervous or sensitive skin, waxy in appearance, called the ceroma.

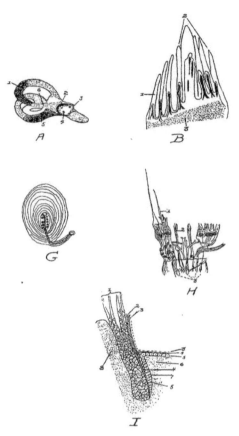

FIG. 29.—Various anatomical parts of the fowl.

A. 1, The superior semicircular canal. 2, The vestibular nerve. 3, The meatus auditorius internus—the entrance of the auditory nerve. 4, The entrance of the cochlear nerve. 5, Posterior semicircular canal. 6, The ampulla.

B. Side papilla of the tongue. 1 and 2, Papillæ showing a fibrous central core supporting blood-vessels and nerves. 3, The basement membrane consisting of connective tissue.

G. Herbst's corpuscle of the tongue of a duck, showing the capsule, lamella and around the nerve zone, numerous round bodies arranged in two rows.

H. From the cristæ acoustica of the ampulla of the dove. A, A vertical section with isolated fiber cells. 1 and 2, The nerve fibers. 3, The hair cells.

I. A longitudinal section through a feather papilla showing a young feather. 1, The developing feather. 2, The horny sheath. 3, The epithelium. 4, Stratum corneum. 5, Stratum malpighi. 6, Corium. 7, Malpighian cell group of the follicle.

The Hard Palate.—The hard palate, palatum durum, forming the roof of the mouth has for its boundaries the beak anteriorly, the mandibular arches laterally, and the pharynx posteriorly. It has four or five transverse bars, projecting posteriorly from each of which is a row of filiform papillæ (Figs. 27 and 29).

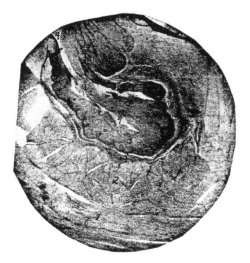

FIG. 30.—A section through the wall of the crop of a hen. 1, The outer muscular layer showing longitudinal fibers. 2, The inner muscular layer showing the sectioned ends of the bundles of muscle fibers. 3, Stratified squamous epithelium. 4, The outer surface or lumen of a fold. 5, The stroma, muscularis mucosa, and submucosa.

The *beak* (Fig. 4, No. 1) with little if any aid from the tongue, is the prehensile organ.

The Tongue (Fig. 27, No. 3). *Location.*—The tongue (lingua) of the fowl is situated in the floor of the mouth between the rami of the lower mandible, and is slung to the cranium by the cerato-branchial element of the os hyoideum (Fig. 18, No. 2).

Shape.—The tongue, is pointed in front and wide behind, shaped like an Indian arrow-head, and is supported by the bony and cartilaginous hyoid apparatus (Fig. 9, *A*).

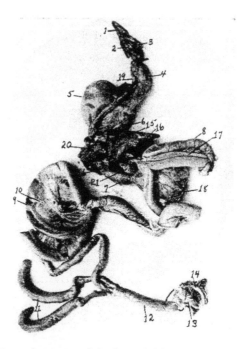

FIG. 31.—The visceral organs of the hen. 1, Tongue. 2, Larynx. 3, Glottis. 4, First portion of the esophagus. 5, Crop. 6, Second portion of the esophagus. 7, Proventriculus. 8, Duodenum. 9, Free or floating portion of the small intestine. 10, Mesentery supporting the free portion of the small intestine. 11, Cæca. 12, Rectum or large intestine. 13, Cloaca. 14, Anus. 15, Liver. 16, Gallbladder. 17, Pancreas. 18, Gizzard. 19, Trachea. 20, Lungs. 21, Spleen.

Structure.—The free part of the tongue consists of two long tubes which are formed by the rolling in of the ventral sides of its horny coverings. The *dorsal plate*, which constitutes one-half of the horny sheath, does not form a part of the tubes, but ends in the region of the point of the entoglossum.

The top surface is covered by a thick stratum corneum, giving it a rather horny surface. Glands occur in the posterior part of the tongue.

While not so freely movable as in mammals, the tongues of birds are very flexible.

The Pharynx (Fig. 27). *Location.*—The anterior part of the pharynx joins with the mouth and the posterior part with the upper portion of the esophagus and the superior larynx. The pharynx is bounded anteriorly by the base of the tongue and by the posterior edge of the hard palate. Posteriorly, it is marked by the entrance of the esophagus and by a ridge of filiform papillæ located on the supero-posterior part of the larynx, called the palatal papillaries. A few filiform papillæ stud the roof. The pharynx communicates anteriorly with the posterior nares and the mouth, and, posteriorly, with the esophagus and the larynx.

Shape.—The pharynx is a musculo-membranous sac.

Structure.—The pharynx is lined with a mucous membrane covered by squamous epithelium. The mucous membrane is thrown into irregular folds. The bird has two Eustachian tubes. Pneumatic apertures conduct the air from the Eustachian tubes to the pericranial diploë.

In the posterior pharyngeal roof is situated the infundibular crevice.

Function.—The function of the pharynx is to give passage for the air from the posterior nares to the larynx, and to give passage for the food from the *mouth* to the esophagus.

Glands Adjacent to the Mouth and to the Pharynx.—The mucous membrane adjacent to the glottis and in the roof of the mouth contains alveolar glands. The *angular gland* of the mouth, located beneath the zygomatic arch, by some anatomists is considered the rudimentary parotid gland. The *sublingual glands*, or glandulæ sublinguales, are well developed (Fig. 17, No. B, 4). They form conical masses, with the apex directed anteriorly, and occupy a portion of the intermaxillary space. Several ducts from these glands open into the mouth cavity. In the palatine region there are also located glands called the *palatine glands*, which open by many stomata upon the surface.

The Esophagus (Fig. 18, No. 6). *Location and Shape.*—The esophagus is a musculo-membranous tube, capable of great distension, which extends from the pharynx to the proventriculus (Fig. 35, No. 1, and 3).

The esophagus communicates anteriorly with the pharynx. It extends down the neck, lying supero-laterally with regard to the trachea and toward the right side. It enters the thorax above the trachea, and, passing through between the bronchi, terminates into the proventriculus. At the entrance of the thorax and just to the right of the median line, it expands into the crop which divides it into two portions, designated as the first and the second. The first portion of the esophagus is the longer, the length depending upon the size and kind of fowl.

Structure.—The wall is composed of four coats, the mucous membrane, the submucosa, the muscular coats, and the fibrous sheath or tunic. The mucous membrane is pale and is covered with stratified squamous epithelium.

The mucous membrane is loosely attached to the muscular coat by the submucosa. Except during deglutition it lies in longitudinal folds which obliterate the lumen. Opening from the mucosa are lenticular glands, which may be seen on inflating an esophagus and looking through its transparent walls.

The muscular coat of the esophagus is divided into two layers, an outer longitudinal and an inner circular layer.

The outer fibrous sheath connects the esophagus loosely to the surrounding structures.

Function.—The function of the esophagus is to give passage way for the food from the pharynx to the crop and from the crop to the proventriculus.

The Crop (Fig. 35, No. 2; Figs. 30 and 31). *Location and Shape.*—The crop, or ingluvies, saccular in shape, is located at the entrance of the thorax and just to the right of the median line. The first portion of the esophagus empties into the crop superiorly and inferiorly the crop opens into the second portion of the esophagus. Like the esophagus the crop is capable of great distension.

Structure.—The wall of the crop is composed of four coats, the mucous membrane, submucosa, muscular and the outer fibrous.

The crop is lined with mucous membrane containing mucous glands which secrete a mucus to keep the surface moist. The surface of the mucous membrane is covered by stratified squamous epithelium.

The submucosa connects the mucous membrane to the muscular coat.

The wall of the crop is provided with strong muscles. The fibrous coat, or tunic, connects it with the surrounding structures.

Function.—The crop is a storehouse for the food during the hours of feeding, the food when needed by the stomach being gradually discharged from the crop by the contraction of its muscular walls. The subcutaneous cervical muscles which cover this reservoir aid in this discharge.

In the *act of deglutition*, the food, after being subjected to the fluid supplied in the mouth by the adjacent glands, is poised upon the tongue and swallowed partly by a sudden jerk of the head, and partly by means of the pressure of the tongue against the hard palate, the food then passes down the esophagus and lodges first in the crop, till needed by the stomach, when it is passed through the second portion of the esophagus to the first portion of the stomach, the proventriculus. The time during which food remains in the crop depends upon the nature of the food. Animal food will, in part, be retained about eight hours and vegetable foods may not all be passed on for from sixteen to eighteen hours.

The Stomach (Fig. 31, No. 7; Fig. 32, B).—The stomach, or ventriculus, of fowls is made up of two portions, namely, the pars glandularis, or proventriculus, and the pars muscularis, or gizzard.

The Proventriculus. *Location.*—The proventriculus lies in the superior part of the groove formed by the two lobes of the liver, is inferior to the aorta, and is directed slightly to the left, communicating anteriorly with the second portion of the esophagus and posteriorly emptying into the pars muscularis, the gizzard.

Shape.—The proventriculus is round transversely and elongated, in fact, nearly fusiform. In the hen of average size it measures about 1.62 inches long and 0.8 inch in diameter.

Structure.—The wall of the proventriculus has four coats, the mucous, submucous, muscular, and serous.

The inner mucous coat which is raised in folds, is lined with columnar epithelial cells. The mucous membrane contains lymphoid tissue. The mucous coat throughout contains simple tubular glands which secrete a highly acid fluid which finds its way to the surface through small cylindrical ducts lying at right angles to the inner surface of the mucous membrane (Fig. 32, No. B, 1).

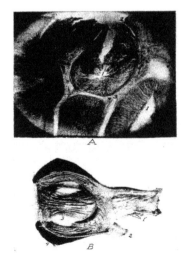

FIG. 32.—Gizzard and stomach of the fowl.

A. 1, Alveolar glands. 2, Mucous membrane of the inner surface of the proventriculus showing the tubular glands. 3, Connective tissue between the alveolar glands. 4, An artery. 5, Muscularis mucosa.

B. Photograph of the proventriculus and gizzard. 1, Proventriculus showing stomata of glands. 2, End of duodenum. 3, Gizzard showing hornified epithelium formed into grooves. 4, The heavy muscular walls.

The submucosa, connects the mucous and muscular coats and the muscularis layer throws the mucous membrane into folds.

Outside the muscularis mucosa there is another layer (Fig. 32, *A*), of simple tubular glands, grouped in lobules, and lined by cuboidal cells and separated from each other by clefts. These tubular glands converge toward the center and open into the same cavity.

The middle muscular coat can be divided into three layers: two thin, longitudinal layers, and a thick, circular layer interposed between the other two.

Function.—The function of the proventriculus appears to be to soak the food with a secretion. The secretion from the proventriculus is similar to that produced by the fundus glands of the stomach of mammals. It contains acid and a ferment-pepsin.

The Gizzard (Fig. 31, No. 18). *Location.*—The gizzard, or muscular stomach, occupies a portion of the central part of the abdominal cavity (Fig. 33, No. 4). It lies slightly to the left and just behind the liver, the proventriculus, and the spleen, and rests upon a mass of intestines. The gizzard communicates at its anterior portion with the proventriculus and with the duodenum. These openings are close together. The gizzards of a large number of hens of average size averaged in weight as follows: full, 0.215 pound; empty, 0.126 pound. The gizzard stands perpendicularly and somewhat obliquely in the abdominal cavity.

Shape.—In shape the gizzard is roundish, flattened laterally.

Structure.—The walls of the gizzard are very thick and are made up of three coats: mucous, muscular, and serous.

The cavity of the gizzard is covered by a thick skin-like structure possessing a heavy stratified squamous epithelial layer which is thrown somewhat into ridges (Fig. 32, *B*). This membrane becomes thinnest near the edges.

The mucous membrane, being cornified and readily detachable, is by some anatomists considered as a special membrane (Fig. 37, *B*).

At the pyloric opening there is a valve formed by a fold of the mucous membrane, which prevents grit and large particles of food from passing out of the gizzard.

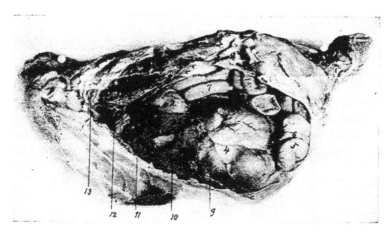

FIG. 33.—View of viscera of the left side of a hen. 1, The base of the heart. 2, Proventriculus. 3, Left lobe of the liver. 4, Gizzard. 5, End of the duodenal loop. 6, Pancreas. 7, Free portion of the small intestine. 8, Lungs. 9, Diaphragm.

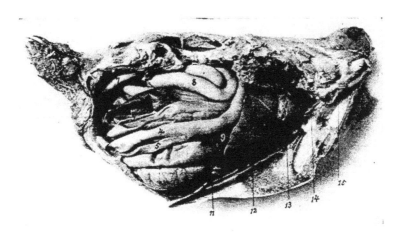

FIG. 34.—View of the viscera of the right side of a hen. 1, Base of the heart. 2, Lungs. 3, Right lobe of the liver. 4, Duodenal loop. 5, Pancreas. 6, Cæcum. 7, Large intestine or rectum. 8, Free portion of the floating small intestine. 9, Gall-bladder. 10, Right kidney. 11, The ribs. Note the lungs pushing up between them.

On each side it has a powerful fleshy muscle, the *muscularis lateralis*. These musculares laterales are hemispherical, consist of very closely packed fibers extending transversely, and are attached to strong anterior and posterior ligamentous tendons. They are joined at the edge of the organ by a strong aponeurosis. The muscular fibers are red but do not possess cross striations; they are of the smooth type. The *muscularis intermedii*, thinner and less developed than the musculares laterales, occurs on each of the anterior and the posterior parts of the gizzard.

The serous coat covers over the greater part of the external surface of the gizzard and closely adheres to the muscular coat.

Function.—The hard callous pads of the gizzard, operated by the powerful muscles above described, together with grit, act like mill stones and make reduction to fineness very complete.

At the posterior part there is a sacculated portion containing glands of the long tubular type which secrete a fluid ferment similar to that secreted by the glands of the pyloric portion of the stomach of mammals. These glands also exist in a small band near the entrance of the gizzard. Here the food is mixed with strongly acid secretion containing pepsin which makes gastric digestion perfect.

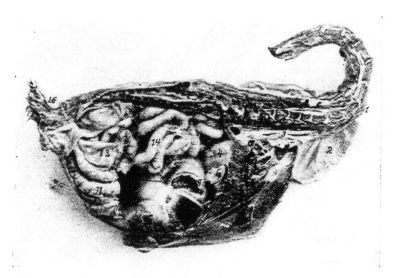

FIG. 35.—A median antero-posterior section through the body of a 1-pound pullet. 1, First portion of the esophagus. 2, Crop. 3, Second portion of the intestine. 4, Proventriculus. 5, Gizzard. 6, Spleen. 7, Liver. 8, Heart. 9, Point where duodenum was severed from gizzard. 10, Point where duodenum was

severed. 11, Duodenum. 12, Pancreas. 13, Cæcum. 14, Floating small intestine. 15, Ovary. 16, Oil sac.

The Small Intestine (Fig. 31, No. 8 and 9).—The small intestine, intestinum tenue, is the tube which connects the gizzard with the large intestine. It is divided into two parts, the duodenum and the free portion. Of the three parts as considered in the mammalian intestine, only the first, the *duodenum*, can be distinguished. There is no demarcation between the jejunum and the ileum. The jejunum and the ileum or that part which represents these sections, are in coils suspended from the free border of the mesentery the other border of which is, in turn, attached to the dorsal wall (Fig. 64). The length of the small intestine in a hen of average size is about 61.7 inches.

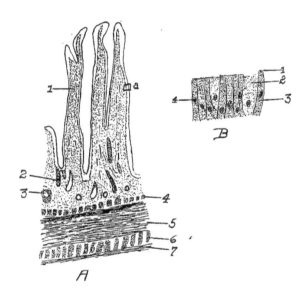

FIG. 36.

A. Section of the duodenum of the fowl. 1, Villus. 2, Gland. 3, Mass of lymphoid tissue. 4, Muscularis mucosa. 5, Longitudinal layer. 6, Circular muscular layer. 7, Serous layer.

B. A section from A at *a*. 1, The striated free border of the cells. 2, Goblet cell. 3, Columnar cell. 4, Nucleus of cell.

The wall of the small intestine is provided with four coats, as follows: a mucous, submucous, muscular coat made up of two layers—the outer, longitudinal and the inner circular layer, and an external serous.

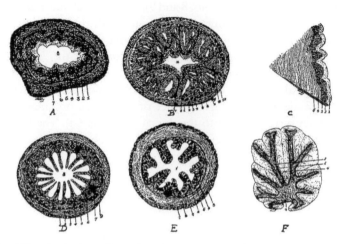

FIG. 37.—Histological studies of various anatomical parts.

A. A transverse section of the first portion of the esophagus of a fowl. 1, The outer longitudinal muscular layer. 2, The circular muscular layer. 3, The submucosa. 4, The muscularis mucosa. 5, Stroma. 6, Epithelial layer. 7, The lenticular glands. 8, The lumen.

B. A transverse section of the proventriculus. 1, The outer longitudinal muscular layer. 2, The middle muscular layer. 3, The inner longitudinal muscular layer. 4, Stroma. 5, The muscularis mucosa. 6, The submucosa. 7, Stroma. 8, Tubular glands. 9, Tubulo-alveolar glands. 10, A tubulo-alveolar gland with the tubular glands cut transversely, 11, The lumen of the proventriculus.

C. A section of the inner wall of the gizzard. 1, The hyaline mucous membrane. 2, Branched tubular glands. 3, Submucosa. 4, Muscle. 5, A connective tissue septum.

D. A transverse section of the small intestines. 1, The outer longitudinal muscular layer. 2, The inner circular muscular layer. 3, The muscularis mucosa. 4, The stroma. 5, Brunner's glands. 6, A villus. 7, Mass of lymphoid tissue.

E. A transverse section through the cecum. 1, The outer longitudinal muscular layer. 2, The inner circular muscular layer. 3, The submucosa. 4, The muscularis mucosa. 5, The stroma. 6, Tubular glands. 7, The lumen.

F. The cerebellum showing the arbo-vitæ. 1, White, fiber portion. 2, The granular layer. 3, The layer of Purkinje cells. 4, The molecular layer.

The mucous membrane which lines the intestine is thick, soft, and highly vascular. It has a velvety appearance, due to numerous long, thin projecting *villi*. The villi (Fig. 36, *A*) are concerned in the absorption of the digested food, absorbing principally the emulsified fats. Each villus is covered with a single layer of high columnar epithelial cells. Some of these, the so-called goblet cells, provide mucin which lubricates the mucous surface. These cells are found in all mucous surfaces and prevent the surface from becoming dry. There are a few goblet cells near the summits of the villi. Openings of simple intestinal tubular glands the duodenal glands, or the glands formerly known as *Brunner's glands*, are located between the villi. These glands secrete the succus entericus, or intestinal juice. These openings, or stomata, are lined with granular cells. The reaction of the contents of the small intestine is strongly acid, but gradually less so in proportion to the distance down the intestinal tract until the cæca are reached, where the reaction is found to be faintly acid, neutral, or slightly alkaline.

Function.—The function of the small intestine is that of digestion and absorption.

The Duodenum. *Location and Shape.*—The duodenum, a small tube, originates from the gizzard about ½ inch to the right of the entrance of the proventriculus. Extending from left to right, it passes under and behind the gizzard along the inner side of the right abdominal wall backward to the posterior portion of the wall, and a trifle more than half way toward the left side, where the loop rests. This loop, the duodenal loop, is about 5 inches in length (Fig. 31, No. 8). The two branches of the loop the first and the second portions or the descending and ascending limbs, are loosely held by connective tissue, and have the pancreas lodged between them.

The Free Portion of the Small Intestine. *Location.*—That section of the small intestine following the duodenum is called the free portion of the small intestine and occupies the space between the abdominal air-sac and the median line of the abdominal cavity. It is disposed in coils and is suspended from the dorsal wall of the abdomen by a thin membrane, the mesentery.

The bile ducts enter the small intestine about 14 inches from its point of origin. The pancreas also pours its contents into the small intestine.

The Large Intestine (Fig. 31, No. 12). *Location.*—The large intestine extends in a straight line along the inferior border of the vertebral column communicating anteriorly with the small intestine, and the cæca and posteriorly with the cloaca.

Shape.—The diameter of the large intestine is approximately twice that of the small intestine. In fowls of average size its mean length is 4.61 inches. The large intestine has sometimes been spoken of as the rectum, or straight gut.

Structure.—Like the small intestine the large intestine has four coats: an inner mucous, a submucous, a middle muscular and an outer serous. The folds of mucous membrane of the large intestine have tubular glands lined with columnar cells (Fig. 37, *A*, No. 1).

Function.—The large intestine is similar in function to that of the small intestine, in that digestion and absorption may take place within it.

The Cæca (Fig. 31, No. 11). *Location.*—The fowl has two cæca extending forward from their point of origin at the juncture of the small and large intestine.

Shape.—The cæca average 7.61 inches in length. They are large in caliber toward the blind extremity and are constricted near their origin.

Structure.—The parietal coats are continued from the small intestine.

Ebarth has described an elevated body in the cæcum, which is located about 4 millimeters from the opening and is composed entirely of lymphoid tissue.

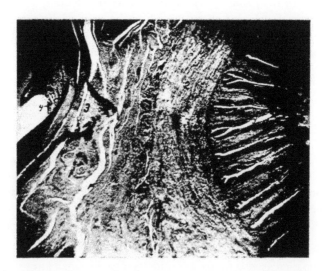

FIG. 37*A*.—Photomicrograph of a transverse section of large intestine and oviduct just anterior to the cloaca, showing the tubular glands of the large

intestines, mucosa of oviduct with intervening structures. 1, The tubular glands of large intestine, or rectum. 2, Muscular wall of intestine. 3, Wall of the oviduct. 4, Mucosa of the oviduct.

The cæca are usually partly filled with a soft pultaceous material of a pasty consistency.

The Cloaca (Fig. 31, No. 13). *Location.*—The rectum opens by a circular, valvular orifice into the dilated portion just in front of the anus, which dilatation is called the cloaca.

Shape.—The cloaca is saccular in shape.

Structure.—The cloaca is divided into two portions, the coprodeumal and the urodeumal. That portion of the cloaca into which the intestine empties is called the *coprodeum*; and the ureter and oviduct empty into the *urodeum*. The seminiferous tubules, carrying the semen from the testes in the male empty in teatlike projections on the cloacal mucous membrane into the urodeumal portion. The cloacal walls are similar in structure to the large intestines.

On the dorsal wall of the cloaca between it and the spine, is a small sac, called the *bursa of Fabricius*, which has a duct communicating with the cloaca. The mucous membrane of this sac is thrown into folds and is studded with glands. The bursa of Fabricius is larger in the young than in the adult bird. It apparently atrophies as the bird becomes older. When the bird is four months old this bursa is best developed, and at this age it may be as large as 2 or 3 centimeters in diameter.

Function.—The function of the cloaca is to give passage way to the feces, the urine, and the egg and to act as an organ of copulation.

Course of the Food.—The food first enters the mouth, after being picked up by aid of the beak. From here it passes through the pharynx and first portion of the esophagus to the crop, without mastication, as the bird is not provided with teeth. The food is passed from the crop by aid of its muscular walls as needed; thence through the second portion of the esophagus to the proventriculus, an expansion in the digestive tube just before it terminates in the gizzard. The glands of the proventriculus produce a secretion in which the food is soaked before passing into the gizzard. The gizzard is provided with strong muscular walls which, by aid of grit, thoroughly reduce the food to fineness. From the gizzard the food passes through the first portion of the small intestine, where it is subjected to the action of the bile from the liver, the pancreatic juice from the pancreas, and of the succus entericus from the glands of the intestinal wall. The food is then passed into the cæca. The indigestible portion of the food passes from the cæca through the large intestine, or rectum, to the cloaca and thence to the external world.

The digestive functions of the bird are very potent and rapid. This compensates for the waste caused by their extensive, frequent, and energetic motions, and is in accordance with the rapidity of their circulation and their high state of irritability.

THE ACCESSORY ORGANS OF DIGESTION

The accessory organs of digestion are the liver, pancreas, and some anatomists include also the spleen. The first two manufacture fluids containing ferments which aid in splitting or digesting the food.

The Liver (Figs. 31, 33 and 34). *Location.*—The liver, hepar, lies ventrally and posteriorly to the heart. It is related anteriorly with the diaphragm, inferiorly with the sternum, posteriorly with the gizzard and intestine, and superiorly with the ovary, oviduct and proventriculus and laterally with the abdominal wall.

Shape.—The liver is a voluminous deep livid brown gland, soft and friable in texture. It is divided into two principal lobes, a right and a left.

The right lobe is larger than the left. In the hen of average size the liver weighs 35 grams. The parietal surface is convex and smooth. The surface which lies against the viscera is irregularly concave. The visceral surface furnishes exit for the bile duct and passage for the nerves and blood-vessels. This part is called the porta.

The left lobe may be cleft from below so deeply as to form two lobes on that side.

Structure.—Each lobe is covered by a double serous membrane, one closely adherent, the other surrounding the structure loosely. These tunics, which are reflections of the peritoneum, are continued from the base of the liver, over both the anterior, and the posterior surface. The loose layer is formed by the air cells surrounding the lobes. The thin border of the liver is usually free.

The two lobes of the liver are connected by a narrow isthmus of liver tissue. Occasionally there is a bird in which there occurs a *lobus Spigelii* located at the posterior of the liver between the two principal lobes.

The apex of the heart sacculates the diaphragm backward, so that part of this apex lies between the right and the left lobes (Fig. 43, No. 6 and 7). A ligament, the *falciform*, extends from the apex of the pericardial membrane, and attaching it rather firmly to the central connective tissue, or interlobar ligament. This ligament also has attachments to the inner surface of the sternum. The *broad ligament* of the liver is formed posteriorly by a fold of the peritoneum.

The *interlobar*, or *principal, ligament of the liver* is formed by a large and strong duplicature of the peritoneum, which makes a longitudinal division in the abdominal cavity similar to the lateral division made by the thoracic mediastinum in mammals. It is reflected upon the pericardium from the linea alba and the middle line of the sternum, and passes deeply into the interspace of the lobes of the liver. It is attached to these lobes throughout their whole length and connects them below to one side of the gizzard. The lateral and posterior part of the liver attach to the adjacent air cells, and the whole viscus is thus kept fixed in its position during rapid and violent movements of the bird.

The remains of the umbilical veins are traceable within the duplicature of the membranes forming the septum. These remains thus represent the round ligament of mammals.

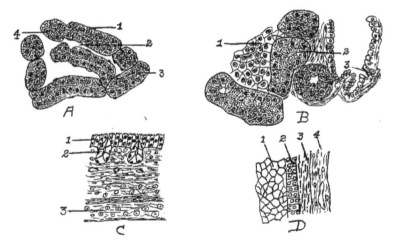

FIG. 38.—Cellular structure of liver, pancreas, and trachea.

A. Liver. 1, Liver cells. 2, Sinusoid. 4, Nucleus of cells.

B. Pancreas. 1, Island of Langerhans. 2, Alveolar cells. 3, Duct.

C. Trachea. 1, Ciliated epithelia. 2, Glands. 3, Hyaline cartilage.

D. Section of wall of ovum at Fig. 57 letter *d*. 1, Yolk. 2, Granular membrane. 3, Theca. 4, Blood-vessel.

A microscopic study of the liver of the fowl shows a compact mass of liver cells polyhedral in shape, with large nuclei (Fig. 38, *A*). The liver tissue differs from that of mammals in that there is no clearly outlined lobular arrangement; neither is the outline of the individual cell so well marked. The parenchymatous portion is made up of columns of liver tissue. These columns anastomose and show narrow channels between. They are best seen in the young chick.

Function.—One of the functions of the liver is to secrete bile. A *gall-bladder*, which receives part of the bile secreted by the right lobe of the liver, is located on the posterior face of this lobe. Extending from this gall-bladder there is a duct, the *cystic duct*, which empties into the small intestine toward the extremity of the second branch of the duodenal loop. Another duct called the *hepatic duct*, proceeds directly from the two lobes of the liver and empties into the intestine just in front of the cystic duct.

The Pancreas (Fig. 31, No. 17). *Location and Shape.*—The pancreas, an organ lying in the duodenal loop is a yellowish-white, lobulated gland, elongated in shape. Its average length in fowls of average size is 4.96 inches, and the average weight about 0.008 pound. The pancreas is divided into lobes, which in turn are divided into lobules.

Structure.—The pancreas has a supporting connective tissue. The lobules are made up of small alveolar glands, which are lined with columnar epithelial cells. The ducts leading from the alveoli are small; these unite to form larger ducts in which the epithelium is taller. Between the alveoli throughout the pancreas there are clusters of polyhedral cells which form the *islands of Langerhans*. These islands are said to produce an internal secretion, or hormone, which is absorbed by the blood or lymph capillaries and thus enters the circulation (Fig. 38, *B*).

Function.—The function of the pancreas is to secrete a fluid containing digestive ferments.

The Spleen (Fig. 31, No. 21). *Location.*—The spleen, or lien, lies in a triangle formed by the proventriculus, the liver, and the gizzard.

Shape.—It is a reddish-brown body shaped like a buckeye, is small in size, weighing only about 0.005 pound.

Structure.—The outer surface of the spleen is covered by a reflection of the peritoneum. After this covering is removed, there is observed a firm, white, fibrous layer, the *cortical portion*. This covering sends into the interior small and large trabeculæ, forming a framework and dividing it into acini, or compartments. The spaces are filled with a dark red parenchymatous material called *splenic pulp*. The framework, in both deep and surface portions, are found elastic fibers and smooth muscular fibers.

The spleen is essentially a lymphatic organ, its peculiar structure depending largely upon the arrangement of the blood-vessels. Compact lymphatic tissue occurs in the spleen in spherical, oval, or cylindrical collections of closely packed lymphoid cells. These masses are known as the *Malpighian bodies*, or *splenic corpuscles*. They are distributed throughout the splenic pulp. Each splenic corpuscle contains one or more small arteries. These extend near the periphery of the corpuscle and more rarely in the center.

The splenic artery passes in and the splenic veins out at the hilum which is located on the concave, or attached, side of the spleen. The splenic artery, upon entering the organ, at once branches, the trabeculæ forming a support for the vessels.

After the arteries have entered the hilum, as stated above, they divide into many branches which follow the septa of the connective tissue. At first the arteries are accompanied by branches of the splenic veins. Soon, however, the arteries leave the veins and the septa, and penetrate the splenic pulp. In the splenic pulp the adventitia of the smaller arteries assume the character of reticular tissue, and become infiltrated with lymphoid cells. This infiltration forms masses which are called the splenic corpuscles, or Malpighian bodies. The terminal arteries break up into capillaries, which still retain an adventitia, and empty into border spaces, or sinuses, sometimes spoken of as ampullæ. These sinuses in turn empty into cavernous sinuses of the splenic pulp. From these are finally formed the venules; and the collections of venules form the splenic veins through which the blood gains exit from the spleen.

The Abdominal and Pelvic Cavities.—In birds the abdominal cavity is divided into two smaller cavities by a fibrous septum. The anterior cavity representing the abdominal contains the liver, and the other representing the pelvic contains the gizzard, intestines and oviduct.

The Peritoneum and the Mesentery.—The abdominal and pelvic cavities are lined by the peritoneum. Like all serous membranes this is composed of a parietal and a visceral portion, which together form a complete sac, with the organs it covers situated on the outer side. The peritoneum like other serous membranes consists of a mesothelial and a submesothelial portion, the cells of the former being arranged in a single layer. Since a serous membrane is so arranged as to line a closed cavity, and at the same time to cover its contents, it follows that the entire membrane must be a closed sac, the mesothelial layer being on the inside; such a sac is called a serous sac, or cavity. Synovial membranes are also regarded as a variety of serous membranes. The fold, or layer, of the membrane which lines the cavity is called the parietal, that which covers the greater part of the organs contained therein is called the visceral portion; the two surfaces contacting, and gliding upon each other, are lubricated by a fluid secretion contained in the sac;

hence one use of these membranes is to prevent friction between the walls of cavities and the organs contained therein. Serous membranes line the abdominal cavity, pericardium, cavities of the heart and is continuous throughout the vascular structures.

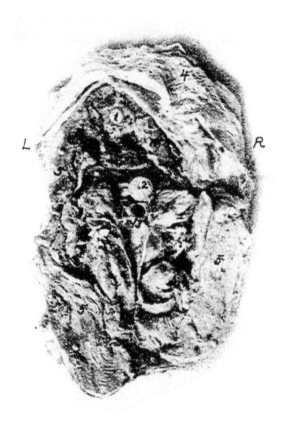

FIG. 39.—A transverse section of body of hen through 15, Fig. 34. R, Right side. L, Left side. 1, Spinal cord. 2, Esophagus. 3, Trachea. 4, Skin. 5, Pectoral muscles.

The serous membrane besides covering the external surface of the viscera, double folds pass from one organ to another, or from an organ to the parietes of the cavity. These double folds of the peritoneum are known as ligaments, or as mesenteries. In ligaments the two folds are strengthened by an interposed layer of fibro-elastic tissue. A *mesentery* is a broad, double fold of peritoneum, attached to the abdominal parietes above, and containing a portion of the intestine in its free or remote extremity. Between its folds we find blood-vessels, nerves, and lymphatics or lacteals, hence it permits vascular and nervous communications with the organ attached to it. The free portion of the small intestine is attached to the free margin of the mesentery.

THE RELATIONS OF THE VISCERAL ORGANS OF THE DOMESTIC FOWL

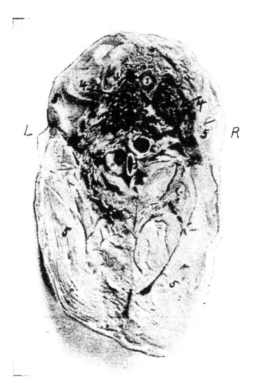

FIG. 40.—Transverse section through the body of a hen through 14, Fig. 34. R, Right side. L, Left side. 1, Spinal cord. 2, Esophagus. 3, Trachea near inferior larynx. 4, Lungs. 5, Pectoral muscles.

Figure 33 shows a fowl with the left abdominal wall and the left thoracic wall removed. No. 1 in this figure shows the base of the heart in front of the left lobe of the liver. Above the liver is the proventriculus; above this, the diaphragm; and above the diaphragm and the base of the heart the left lung occupying the superior part of the thoracic cavity, and that there is no distinct pleural sac, as in mammals but that the lung pushes out between the ribs, thus pressing against the ribs on the inner and the lateral sides. The gizzard is back of the liver, to the left side of the abdominal cavity, and beneath and in front of the duodenal loop. The small intestine from this side is above the gizzard. Supero-anterior to the gizzard is the blind end of the cæcum. The pancreas is within the duodenal loop.

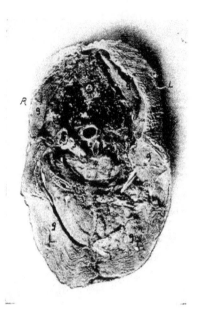

FIG. 41.—Transverse section through the body of a hen at 12, Fig. 33. R, Right side. L, Left side. 1, Spinal cord. 2, Esophagus. 3, Inferior larynx. 4, Base of the heart. 5, Aorta. 6, Vena cava. 7, Lungs. 8, Skin. 9, Pectoral muscles.

Figure 34 shows the viscera from the right side after the removal of the right abdominal and the right thoracic wall. The base of the heart is in front of the right lobe of the liver. Above these the right lung occupies the upper part of the thoracic cavity as in the preceding illustration. Just back of this is the anterior lobe of the kidney. The gall-bladder is observed at No. 9 on the right lobe of the liver. Just inferior to a longitudinal central line is the duodenal loop, between the limbs of which is seen the pancreas. Above this loop the cæca are located. The gizzard is not visible from the right side; on this side posterior to the liver is the small intestine. Superior to this at No. 7 the rectum.

The relative position of the visceral organs in the median line is observed in Fig. 35. No. 1 of this figure shows the stump of the first portion of the esophagus, and No. 2, the left wall of the crop. Following this, and located just below the vertebræ is the second portion of the esophagus. The major portion of the crop is located on the right side.

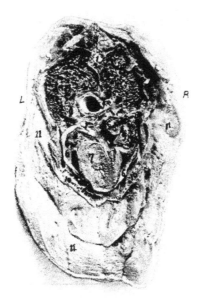

FIG. 42.—Transverse section through the body of a hen at 11, Fig. 33. R, Right side. L, Left side. 1, Spinal cord. 2, Vertebra. 3, Spinous process of vertebra. 4, Lungs. 5, Esophagus. 6, Pericardial sac. 7, Sectioned surface of heart. 8, Auricle of heart. 9, Blood clot in right auricle. 10, Section of sternum. 11, Pectoral muscles.

It will be noted that the second portion of the esophagus passes over the base of the heart and the superior part of the liver, and then terminates in the proventriculus. The proventriculus extends downward and empties into the gizzard. This organ lies antero-laterally to the gizzard, supero-posterior to the liver, and to the left of the spleen. The spleen lies in a triangle formed by the liver, the proventriculus, and the gizzard. The heart is noted to lie supero-anterior to the liver, and between the anterior portion of the fissure formed by the right and the left lobe. The ovary is located back of the diaphragm at the anterior end of the kidney and below the inferior surface of the bodies of the vertebræ. The bulk of the floating portion of the small intestine is located above the gizzard.

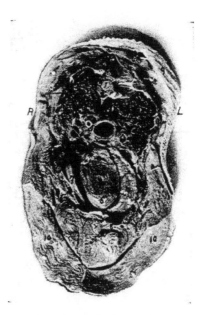

FIG. 43.—Transverse section through the body of a hen at 13, Fig. 34. R, Right side. L, Left side. 1, Spinal cord. 2, Body of vertebra. 3, Superior spinous process of vertebra. 4, Lungs. 5, Esophagus. 6, Heart. 7, Right and left lobes of liver. 8, Sternum. 9, Skin. 10, Pectoral muscles.

A transverse anterior section of a normal laying hen is shown in Fig. 39, the section being made at 15 of Fig. 34. At this point the esophagus lies centrally and above the trachea. Figure 40 shows a transverse anterior section through the thoracic region at 14 of Fig. 34. At this point the esophagus is slightly to the right of and is superior to the trachea. At this level the apex of the lung is sectioned. This is in the region of the cervical air-sac. Figure 41 shows a posterior section at 12 of Fig. 33. The esophagus here is above and to the right of the inferior larynx, and directly below and between the lungs. The inferior larynx is above the base of the heart.

Figure 42 shows a transverse anterior section made at 11 of Fig. 33. It shows that at this level the esophagus is centrally located and passes over the base of the heart. The heart occupies the lower portion of the thorax and the lungs the upper. Figure 43 shows a photograph of a transverse anterior section of the body made at 13 of Fig. 34. At this level the apex of the heart lies within the anterior fissure of the liver. The sectioned portion of the heart shows the lower portion of the ventricle.

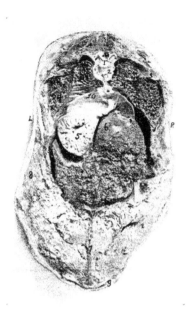

FIG. 44.—Transverse section through the body of a hen at 12, Fig. 34. R, Right side. L, Left side. 1, Spinal cord. 2, Articular surface of vertebral segment. 3, End of rib. 4, Lungs. 5, Proventriculus. 6, Liver showing Glisson's capsule. 7, Sternum. 8, Pectoral muscles. 9, Skin.

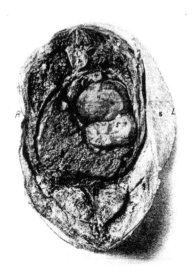

FIG. 45.—Transverse section through the body of a hen at 10, Fig. 33. R, Right side. L, Left side. 1, Spinal cord. 2, Body of vertebra. 3, Lungs. 4, Ova. 5, Proventriculus. 6, Liver. 7, Sternum. 8, Skin. 9, Pectoral muscles.

In this figure the lungs show, on the sectioned surface, the ends of some of the larger bronchi. The esophagus is located above the heart. Showing the relations of the visceral organs back of the heart girdle, photograph number 44, gives an anterior section made at 12, Fig. 33. The lungs are spread out occupying the posterior thoracic region and below at No. 10 is the diaphragm. The diaphragm does not appear to have that rigidity and firmness of position as in mammals. It is rather rudimentary. Below the left lung and above the left lobe of the liver is the proventriculus. Note that the viscus is empty and that the mucous membrane is thrown into folds. At this point the liver occupies much of the abdominal cavity. Figure 45 is a view of an anterior section made at 10, Fig. 33. In this section the lungs are decreasing in caliber. The liver occupies much of the space in the lower right abdominal quadrant, and above and to the right is the ovary with many of the ova developing yolks. Below No. 4 which is a developing yolk, is the proventriculus. Figure 46 is a view of an anterior section made at 9, Fig. 33. At this level are shown sectioned surfaces of the kidneys, which lie on either side of the spinal column. Below the spinal column and occupying the left upper quadrant is the ovary containing ova in the process of developing yolks. To the right is the sectioned surface of one of the cæca and below this and on the right side is the sectioned ends of many of the loops of the floating portion of the small intestine. Occupying the left lower quadrant is the sectioned surface of the gizzard and on the abdominal floor and to the right of the gizzard, the posterior end of the right lobe of the liver.

FIG. 46.—A transverse section through the body of a hen at 9, Fig. 33. *R*, Right side. *L*, Left side. 1, Spinal cord. 2, Vertebra. 3, Kidneys. 4, Ovary. 5,

Cæcum. 6, Small intestine. 7, Gizzard. 8, Right lobe of liver. 9, Skin. 10, Pectoral muscles.

THE RELATIONS OF THE VISCERAL ORGANS OF THE BABY CHICK

There is approximately 47 per cent. of the yolk retained in the abdominal yolk sac of the baby chick at hatching. Figure 47 shows a photograph of a longitudinal section through a baby chick. This figure shows all the posterior portion of the abdominal cavity occupied with abdominal yolk. The abdominal viscera are pushed forward, and as the yolk is gradually absorbed the visceral organs gradually occupy their normal position.

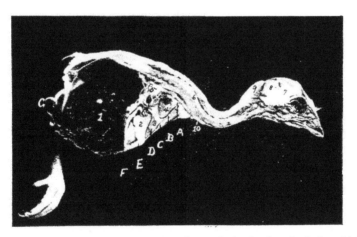

FIG. 47.—An antero-posterior section through the body of a baby chick just hatched. 1, Abdominal yolk sac. 2, Gizzard. 3, Liver. 4, Heart. 5, Intestines. 6, Spinal cord. 7, Cerebrum. 8, Cerebellum. 9, Fat in the post-occipital region. 10, The thymus gland.

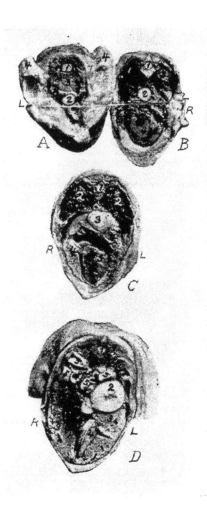

FIG. 48.—Transverse sections of the body of a baby chick at hatching. R, Right side. L, Left side.

A. A transverse section at A, Fig. 50. 1, Spinal cord. 2, Esophagus. 3, Entrance to thorax. 4, Stubs of the wings.

B. A transverse section at B, Fig. 50. 1, Spinal cord. 2, Esophagus. 3, Lungs. 4, Heart.

C. A transverse section through the body at C, Fig. 50. 1, Spinal cord. 2, Lungs. 3, Esophagus. 4, Liver. 5, Heart.

D. A transverse section at D, Fig. 50. 1, Spinal cord. 2, Proventriculus. 3, Liver. 4, Gizzard. 5, Intestine. 6, Kidneys. 7, Gall-bladder.

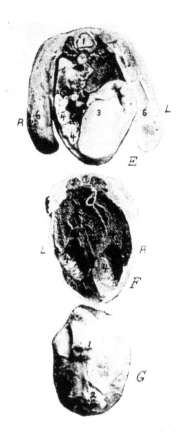

FIG. 49.—Transverse section through the body of a baby chick.

E at E, Fig. 50. 1, Spinal cord. 2, Kidneys. 3, Gizzard. 4, Intestines. 5, Unabsorbed yolk. 6, Stubs of legs.

F. A section at F, Fig. 50. 1, Spinal cord. 2, Kidneys. 3, Intestine. 4, Unabsorbed yolk.

G. A section at G, Fig. 50. 1, Anus. 2, Umbilicus.

A section through this body Fig. 47 at A is shown in Fig. 48, A. At this point the esophagus appears below the vertebral column. Figure 48, B, a section through the body at Fig. 47, B, shows the apex of the lungs. In a median line and below the lungs is the esophagus. Note the mucous membrane thrown into folds. Here the heart is sectioned, showing both auricles and both ventricles. This is a view looking forward. Figure 48, C, is a view of a posterior section made at Fig. 47, C. It shows the same relations of the

- 157 -

esophagus and lungs, but shows at this level both the right and the left lobe of the liver. Between these lobes we note the sectioned apex of the heart. Figure 48, *D* shows a posterior section at Fig. 47, *D*. At this level the kidneys are above and infero-laterally to the spinal column. The intestines are below the kidneys; and on the left side, between the kidneys and left lobe of the liver, and near the abdominal wall, is the proventriculus. Occupying the major portion of the inferior abdominal cavity are the right and the left lobe of the liver, and between these the anterior border of the gizzard. Note the gizzard sitting at an angle inclining toward the left side. Figure 49, *E* shows a section at Fig. 47, *E*. Here the gizzard occupies the left lower abdominal quadrant. To the right are the intestines; and directly above the gizzard is a small portion of the anterior end of the abdominal yolk sac. Infero-laterally are the kidneys. Figure 49, *F*, shows the kidneys similarly located as in the preceding; and just below is the rectum suspended by the mesentery. The rest of the cavity is occupied by the abdominal yolk.

THE URO-GENITAL SYSTEM

The uro-genital apparatus, or apparatus uro-genitalis, consists of two groups of organs: the urinary and the genital. The former elaborate and remove the chief excretory fluid, the urine; and the latter serve for the formation, development, and expulsion of the products of the reproductive glands.

THE URINARY APPARATUS (Fig. 50 and Fig. 51).

The urinary apparatus of the bird consists of two kidneys, from each of which a ureter extends and empties into the cloaca.

The Kidneys. *Location.*—The kidneys are located in excavations in the pelvic roof. They are related internally with the posterior aorta and vena cava, supero-internally with the lumbo-sacral vertebræ and superiorly and supero-externally with the ilium. The abdominal visceral organs lie below the kidneys. The kidneys are external to or above the peritoneum.

Shape.—In the fowl of average size the kidneys are 2½ inches long and are made up of three irregular lobes. The anterior lobe is usually the largest and the middle the smallest. The anterior border of the first lobe is located opposite the last true dorsal articulation. The anterior lobe is called the anterior pelvic or ilio-lumbar lobe, the middle the middle pelvic or ilio-sacral lobe, and the posterior the posterior pelvic lobe.

Structure.—The whole kidney has a fine transparent covering. The dark, brownish-red parenchyme can be seen through this membrane. It has a blood vascular system and a urinary tubular system. The larger arteries, the veins, and the nerves pass between the lobules, and the smaller vessels between the tubules. These form fine network or plexuses. The lymphatic vessels are very few, and are mainly found on the surface. The kidneys are pierced in their posterior third by the external iliac artery and at about its middle by the venous branches forming the posterior vena cava.

The lobes are made up of *lobules*, which are plainly perceptible from the external surface. Each lobule is apparently a unit within itself. It receives its blood supply from small branches of the renal artery and is made up of a cortical or peripheral portion which contains the glomerules (Fig. 52, *A*) and a medullary portion, which is made up of tubules, or urinary canals, arteries, veins, and nerves.

FIG. 50.—Photograph showing spinal cord and relative positions of the kidney, lung and heart. 1, Heart. 2, Lung. Note the indentations made by the ribs. 3, The anterior, middle and posterior lobes of the kidney. The lobules are plainly visible. 4, The lumbo-sacral segment of the spinal cord. 5, Terminal cord filament. 6, The cervico-dorsal segment of the spinal cord. 7, The brachial nerve plexus. 8 and 9, The lumbo-sacral plexus. 8, The ischiadic nerve. 9, The lumbar nerves.

The *urinary canals* are divided into two kinds, namely, the outer and the inner tubular systems. The outer canals, called the *tubulæ uriniferi corticalis*, are very small in caliber and are located in the lobules. The renal artery (Fig. 53, No. *B*, 3) breaking up into arterioles in the kidney, and finally reaching the cortical portion of the lobules, form capillary plexuses in the shape of minute spheres, which are the glomerules (Fig. 53, No. *C*, 3 and *B*, 5). Around each glomerule there is formed a capsule called *Bowman's capsule*, which is the beginning of the uriniferous tubule. This entire mass is called the *Malpighian body*, or *renal corpuscle*. This capsule then extends as the urinary, or secreting tubule, being at first constricted, then convoluted, and terminates into a second portion, the *descending limb of Henle*. This portion of the tubule becomes constricted. It then forms the loop of Henle and ascends as the second portion, or second limb of Henle, which is again of greater diameter. The top of the secreting tubule is slightly wavy and empties into the collecting tubule along with many others. These collecting tubules in turn merge into large tubules which finally empty into the ureter (Fig. 53, No. 14). These collective bundles correspond to the pyramids of the kidneys of mammals.

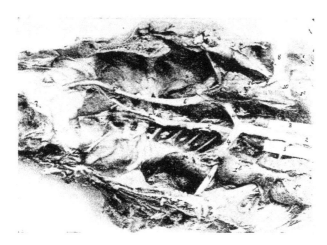

FIG. 51.—The kidneys. 1*a*, The posterior. 1*b*, the middle and 1*c*, the anterior lobes of the kidney. 2, The posterior aorta. 3, The external iliac or crural artery. 4, The ischiadic artery. 5, The sacralis media artery. 6, The ureter which empties into the cloaca at 7. 8, The external iliac vein. 9, The internal iliac vein. 10, The iliacus communis.

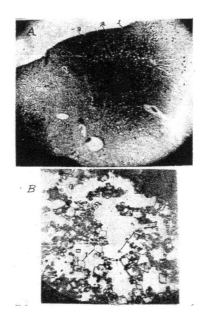

FIG. 52.

A. Photomicrograph of a section of a kidney showing three lobules. 1, Cortical portion of lobule showing glomeruli. 2, The central or medullary portion of the lobule. 3, The outer surface of the kidney.

B. Salts from the urine of a hen. 1, Uric acid crystals. 2, Sodium urate crystals.

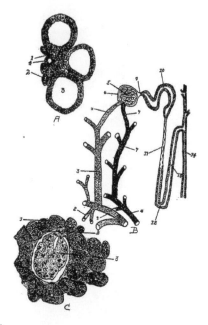

FIG. 53.—The renal structure.

A. 1, Capillary blood-vessel. 2, Descending limb of Henle. 3, Collecting tubule. 4, Ascending limb of Henle.

B. 1, Branch of renal artery. 2, The descending or medullary branch. 3, The ascending branch. 4, The arteriole taking part in the formation of the glomerule. 5, The glomerule. 6, Bowman's capsule. 7, Blood-vessel extending from the glomerule. 8, The vein of the medullary part of the lobule. 9, The neck. 10, The convoluted tubule. 11, The descending limb of Henle. 12, Henle's loop. 13, The ascending loop of Henle. 14, The collecting tubule.

C. 1, Section through a convoluted tubule. 2, Bowman's capsule. 3, The glomerule.

The neck of the uriniferous tubule as it emerges from Bowman's capsule is short and narrow and is lined with a few cuboidal cells. Toward the glomerular end the cells are of a transitional form gradually merging into the flat squamous type peculiar to Bowman's capsule. The first convoluted tubule is lined with irregular cuboidal or pyramidal epithelial cells. The descending limb of Henle's loop is narrow and is lined with a simple layer of flat epithelial cells. In the loop the epithelium changes from the flat type in the descending limb to the cuboidal in the ascending limb. The ascending limb of Henle's loop again becomes broader and is lined with low cuboidal cells. The second convoluted, or tortuous, tubule is lined with low cuboidal cells as are also the collecting tubules. The collecting tubules originate from every part of the internal substance of the lobules, and extending to the gyrations, uniting in the pinniform structure and traversing to the margin of the lobules, following along the uneven surface, infero-laterally and toward the median body line, finally empty into the ureter.

There are many arterial branches given off from the arteria renalis. The arteries which supply the kidneys are given off from the posterior aorta and ischiadic artery. As soon as these arteries enter the kidney they break up into two systems. One system supplies the kidney substance with nourishment in the form of nutrient blood; the second system supplies the glomerules with what may be considered functional blood (Gadow). The second system of arteries branches into the small arteria interlobularis, which pass between the lobules of the kidneys where they give off side branches which penetrate the cortical portion of the lobules and form the capillary plexus, the glomerule.

The arteriole that enters into the structure of the glomerule is lined with endothelial cells and is surrounded by a few muscle fibers and a fine network of connective tissue. The vessel that carries the blood away is similarly constructed.

Function.—In the glomerules the liquid portion of the urine is filtered out of the blood, which urine flows through the uriniferous tubules and passes from the kidney through the ureter. In the cubical cells are extracted the solid portions of the urine which also pass through the tubules with the liquid. The urinary secretion, as found in the ureter, does not contain much liquid, but, on the contrary, is made up of a pasty material consisting of salts which are, for the most part, uric acid crystals and sodium urate (Fig. 52, *B*). This material becomes hard, like cement, soon after being exposed to the atmosphere. This secretion may be noted as a whitish pasty material on the outer parts of the feces, or droppings, voided by the birds.

The Ureter. *Location.*—The ureter extends along the inferior surface of the kidney. It has its origin near the anterior extremity of the kidney, and passing posteriorly the entire length of the kidney receives tributary collecting tubules, and terminates in the upper wall of the cloaca in the urodeumal portion.

Shape.—The ureter gradually enlarges in diameter until it reaches the posterior border of the kidney, and then maintains about the same caliber throughout the rest of its course.

Structure.—The ureter wall is made up of three coats as follows: mucous, muscular and fibrous.

Function.—The ureters serve as a passage way for the urine from the kidneys to the cloaca.

THE MALE GENERATIVE ORGANS (Fig. 54)

The male generative organs in the fowl consist of two testicles and an excretory apparatus, the vas deferens, for each.

The Testicles. *Location.*—The testicles are located in the sublumbar region of the abdominal cavity, behind the lungs, below the anterior extremity of the kidneys, and opposite the last three ribs.

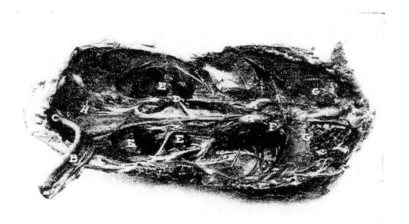

FIG. 54.—The pelvic organs of a cockerel. *A*, Testes. *B*, Rectum. *C*, Cloaca. *D*, Vas deferens. *E*, Kidney. *F*, Adrenal gland. *G*, Lungs. *H*, Ureter.

Shape.—The testicles are ellipsoid in shape; their size varies with the different species of birds. The two in the same bird are usually of the same size, though one testicle may be slightly larger than the other. In a summary of a large number of weights of the testicles of ten months old cockerels, the average weight was 0.021 pound each. The average measurements were 2 inches in the major diameter and 1 inch in the minor diameter. In the cockerel before sexual maturity, which is denoted by the male bird's crowing, the testicles are very small. They resemble, in shape, a navy bean and are yellowish-white in color.

Structure.—The testicle is surrounded by a thin fibrous capsule, which, in the mature cock, is very vascular. This capsule sends into the interior of the gland, septa which form the framework, or supporting structure. This framework forms the spaces in which are located the glandular substance. The glandular portion consists of the *tubuli seminiferi*, which are lined with cubical cells. The framework supporting these tubules gives passage to arterial branches of the spermatic artery, which furnish an abundant blood supply; the framework also supports the veins returning the blood from the testicle. The seminiferous tubules end in blind extremities in the epididymus or globus minor, and unite in the seminiferous canals. All arteries, veins, lymphatics, and seminiferous tubules enter or leave through the globus minor at the attached portion of the testicle. The *epididymis* is made up largely of convoluted tubules which are the continuation of the secreting tubules of the globus major. The walls of the tubules of the globus minor become thicker and are provided with smooth muscle cells in addition to the connective tissue and endothelial lining. The convoluted tubules empty into the vas deferens. The epididymus is covered by the fibrous capsule; which corresponds to the tunica albuginea of mammals and may be considered as a reflection or modification of the visceral peritoneum.

The substance of the testicle is very soft; in fact, it may be said to be of the consistency of encephaloid material. It is made up of secreting tubules in which are found the spermatozoa; it also contains cells which produce an internal secretion, or hormone. There is also produced some fluid in which the spermatozoa float, the whole material manufactured constituting the semen.

The *spermatozoa* of the fowl are provided with long cylindrical bodies, which may be straight or wavy (Fig. 55, *A*). The body of the spermatozoon is obtuse anteriorly, and posteriorly, tapers into a filimentary tail, or flagellum, of varying length, by the aid of which the spermatozoon moves about in the fluid.

The histological structure of the testicles of the baby chick at hatching is approximately one-half white fibrous connective tissue. The seminal tubules are small and widely distributed among the connective tissue. The cells of these tubules possess rather large nuclei, round in shape, with linin network and chromatin granules, typical of resting germ cells. As the bird develops the testicles grow, the seminal tubules become larger, and the amount of connective tissue correspondingly less.

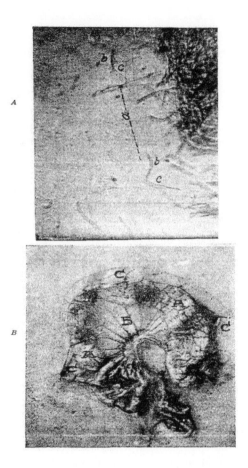

FIG. 55.

A. Spermatozoa of a cock. *a*, The spermatozoa. *b*, The head. *c*, The tail.

B. The oviduct of a hen removed. *A*, The oviduct. *B*, The superior ligament. *C*, The inferior ligament. Note the oviduct thrown in folds and the anastomosing blood-vessels.

In addition to the tubules, the spermatogenic cells, and the connective tissue, there is also in the testicles more or less fat. There is a small amount of connective tissue between the seminiferous tubules, in which locations there are also clusters of polyhedral cells, with round nuclei, others are the interstitial cells.

The Vas Deferens (Fig. 54, *D*). *Location and Shape.*—Extending from the epididymus, is the vas deferens which runs backward on the infero-internal surface of the kidney and to the outside of the ureter. It is very tortuous passing on the infero-lateral surface of the kidney in company with the ureter and becoming somewhat expanded posteriorly it terminates in the upper wall of the cloaca in a rather small papilla located in the uro-genital portion of the cloaca anterior to the mouth of the ureter. This papilla is the organ of copulation and in ducks is very large, and spirally elongated, and retractile forming a kind of penis. The papilla is traversed by a furrow on the upper surface through which the semen flows.

Structure.—The vas deferens is covered and supported by the peritoneum. Its wall is made up of a fibrous structure in which may be found smooth muscle fibers. The wall does not possess glands. It is lined with columnar epithelium. The posterior end is expanded and terminates into a papilla. The base of the papilla is surrounded by a plexus of arteries and veins, which serve as an erectile organ during the venereal orgasm, when the fossa of the turgid papilla is everted, and the semen brought into contact with the similarly everted orifice of the oviduct of the female, along which the spermatozoa pass by undulatory movements of their ciliary appendage, or tail.

THE FEMALE GENERATIVE ORGANS (Fig. 56)

The female generative organs consist of one ovary and an oviduct.

Location.—The ovary is located similarly to the testicles of the male bird, in the sublumbar region of the abdominal cavity, just at the anterior end of the kidneys, posterior to the lungs, and slightly to the left of the center.

Shape.—In the pullet the ovarian mass appears somewhat like a bunch of grapes, being made up of from 3500 to 4500 small, whitish spheres, which represent the undeveloped ova, and which in the active state are developed, one by one, into yolks with their blastoderms. From the blastoderm the fetus may later be developed. In the active ovary of the laying hen the ovarian mass is of considerable size, as it contains ova in different stages of development. Only one ovum is completely developed at a time, though occasionally there may be only a few hours between the maturity of successive ova. The ova receives nourishment from the blood-vessels of the capsule, which vessels are branches of the ovarian artery.

FIG. 56.—Functionating female generative organs of a hen. 1, Ova in process of formation of yolk. 2, Stigmal line at which point the capsule ruptures when ovum is mature. 3, The funnel end of the oviduct. 4, The oviduct torn loose and laid to one side, the albumin-secreting portion. 5, The shell membrane secreting portion. 6, The albumin. 7, The yolk. 8, The shell-secreting portion. 9, The cloaca. 10, The rectum.

Structure.—The ovary contains very vascular cellulofibrous tissue. The ovum as it develops is attached to the ovarian body by means of a delicate white fibrous pedicle. When the yolk is mature it escapes from the enveloping fibrous capsule by a cleavage of the capsule. The cleavage line is called the *stigmen* (Fig. 56, No. 2). The yolk is surrounded by a very delicate membrane called the *vitelline membrane*. The empty sac now shrinks and finally disappears.

The Egg.—The principal divisions of the egg are the yolk, the albumin outside of the yolk content, the shell membranes and the shell (Fig. 58). As stated, the *yolk* is formed in the ovary, leaving the other three portions to be formed in the oviduct.

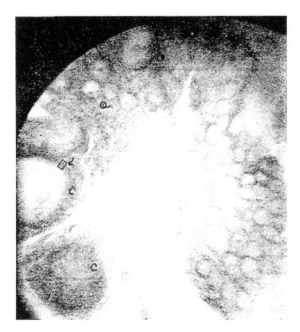

FIG. 57.—Section through the ovary of the hen. *a*, The ova. *d*, An ovum beginning to receive deposits of yolk material. *c*, Ova farther advanced in yolk formation.

The *albumin* may be subdivided into, first, a thin layer of albumin lying close around the yolk; second, a thick layer of albumin lying at the outer periphery; and third, a modification of the albumin called the chalazæ. The chalazæ are twisted, dense cord-like structures at either pole of the yolk, one end of which is adherent to the vitelline membrane and the other to the inner membrane surrounding the albumin. The chalazæ thus act as stays to this structure, which is carrying a delicate burden, the blastoderm.

The *shell membrane* consists of two layers, an inner delicate, and an outer thicker layer. When the egg is just laid these two membranes are in all parts closely adherent to each other, and the egg content completely fills the shell cavity. As soon as the content cools there is a slight contraction; the two shell membranes separate at the large end of the egg, forming an air cell which gradually enlarges as the evaporation of liquid through the pores of the shell takes place.

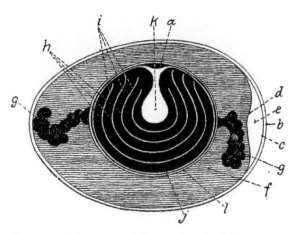

FIG. 58.—A diagram of the parts of the egg. *a*, The blastoderm. *b*, The shell. *c*, The outer shell membrane. *d*, The inner shell membrane. *e*, The air-sac at the large end of the egg. *f*, The albumin. *g*, The chalaza. *h*, The yellow layers of yolk. *i*, The white layers of yolk. *k*, The flask-shaped portion of white yolk. *l*, The vitelline membrane.

The shell consists of several layers. Three are easily distinguishable: first, an inner mammillary layer, consisting of minute conical deposits of calcareous material; second, a middle spongy layer, composed of a thick network of fibers; third, an outer delicate, cuticle-like structure. In certain breeds of poultry a pigment may be added; for example, in ducks a pea green, in turkeys a spotted brownish material, and in fowls pink and various shades of brown. The egg shell is porous to admit the free exchange of air during the process of incubation.

An average sized hen egg weighs about 2 ounces, of which 11 per cent. is shell, 32 per cent. yolk, and 57 per cent. white. The principal chemical constituents of the egg are as follows: ash, or mineral matter, 9 per cent.; fat, or hydrocarbon, 9.3 per cent.; proteids, or nitrogenous matter, 11.9 per cent.; and water, 65.5 per cent. There is apparently no constant proportion of weight between the yolk and albumin. There is also a variation in the weight of the shell due to its variation in thickness.

In an examination of ten eggs of average size, the yolk constituted 31 per cent. of the total weight of the egg.

The following is the result of the analysis of twelve eggs of average size. This analysis included the shell and all other parts taken together.[6]

[6]. The analysis was made by D. M. McCarty, chemist, Animal Industry Division, North Carolina Experiment Station.

Moisture	64.25 per cent.
Dry matter	35.75 per cent.
	100.00 per cent.

Parts per hundred including shell	
Protein	10.2500
Fat	10.6200
Phosphorus	.3020
Calcium	.6080
Magnesium	.0985
Iron	.0103
Sulphur	.3950
Chlorine	.1506
Potassium	.0103
Sodium	.2000

The Oviduct.—The three separate and distinct portions of the egg, albumin, shell membranes, and the shell, are constructed in different parts of the oviduct.

Location.—The oviduct of the hen extends along the left side of the bodies of the vertebræ, and the roof of the pelvic cavity and lies dorsal to the abdominal air-sac. It extends from the posterior border of the ovary and empties into the cloaca (Fig. 56, No. 3 and 8) through a transverse slit.

Shape.—In a well-developed Plymouth Rock pullet, but one whose reproductive organs have never become active, the oviduct is about 4½ inches long; in the fully developed and active state it is from 18 to 20 inches long, and in a collapsed state about ½ inch in diameter. It is held in position by two ligaments, a dorsal and a ventral (Fig. 55, *B*) to be described later. The oviduct is tortuous in its course, forming three principle convolutions before reaching the cloaca, and like the ovarian mass, in its active state, pushes the abdominal viscera downward and toward the right side.

Structure.—The oviduct consists of three coats: first, the serous, located on the outside, which is a reflection of the peritoneum; second, the muscular middle tunic; third, the mucous coat, which in a resting state is thrown into folds.

FIG. 59.—The active oviduct of a hen laid open. 1, The ovary. 2, The funnel. 3, The albumin-secreting portion. 4, The isthmus. 5, The shell=gland portion. 6, The vagina. 7, The superior ligament. 8, The inferior ligament.

Parts of the Oviduct.—The parts of the oviduct are as follows: the funnel, the albumin-secreting portion, the isthmus, the shell-gland portion, or uterus, and the vagina (Fig. 59).

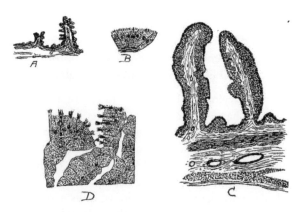

FIG. 60.—The mucous lining of the oviduct.

A. Transverse section of the oviduct wall in the region of the neck of the funnel showing primary and secondary folding of the epithelium (after Surface).

B. Showing the type of gland cells of the funnel region.

C. Transverse section through the wall of the uterus showing the deep folds.

D. Section from the albumin-secreting portion showing the opening of a tubular gland and also showing the character of the cells.

The *funnel* is the trumpet-shaped portion, the ostium tubæ abdominale, whose mouth or fimbriated opening faces the ovary, and lies ventrally to receive the ovum, or yolk, as it is discharged from the ovary. Its thin wall, expanded in the anterior portion, is provided with fimbriæ-like projections. This funnel-shaped portion soon converges to form a constricted portion. This portion of the active oviduct is from 3 to 4 centimeters long. The mucous membrane occurs in folds forming low longitudinal spiral ridges (Fig. 60, *A*). The fimbriæ are continuous with the dorsal and the ventral ligament of the oviduct; and from this point, where the ridges of the mucous membrane are almost nil, they gradually increase in height as they extend down the tube. These ridges are continued in those of the second, or albumin-secreting portion (Fig. 60, *D*). At this point they increase in height very rapidly. Here the bundles of muscular fibers of the middle coat are thin and distributed among bundles of connective tissue. The muscular fibers consist of two layers, an outer longitudinal and an inner circular. At places in this portion the inner bundles may be noted to extend longitudinally. In embryological development the epithelial layer has an origin different from the outer layers of the oviduct. In the fetal development the Müllerian duct arises as a thickening along the Wolffian body just ventral to the gonad. This Müllerian duct is at first a solid cord of cells. It later develops a lumen, and grows posteriorly until it connects with the cloaca. At the time of this posterior growth, mesenchyme cells migrate in from the surrounding tissue and form a layer about the duct. From this layer of mesenchyme cells there are developed the outer layers of the oviduct, which layers later develop the muscular structure and the connective tissue. The epithelium and its derivatives, which represent the glandular structures, are formed from the walls of the old Müllerian duct. Thus the two sets of tissues, having different origins, likewise have different functions. The epithelium is concerned entirely with secretion, and the derivatives of the mesenchyme are concerned with supporting and muscular function.

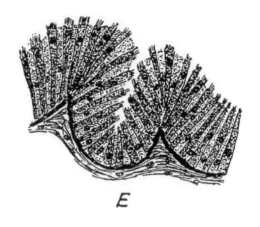

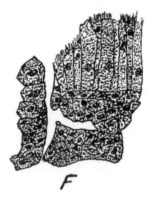

FIG. 60. *A*.—The mucous lining of the oviduct.

E. Section of the epithelium from the vagina showing unicellular glands.

F. A section from the isthmus showing opening of a tubule.

To summarize, six layers of tissue occur in the funnel region, namely, an outer serous covering, an outer longitudinal muscular layer, a layer of connective tissue, an inner circular layer, a second layer of connective tissue, and an inner mucous layer. The mucous layer is made up of glands, as follows. The unicellular glands occur between the ciliated cells of the epithelium. These glands are found only in the posterior half of this division of the oviduct. The glandular grooves are made up of an accumulation of gland cells at the bottom of the grooves between the secondary folds of the epithelium. These are found in all parts of this division except the extreme posterior part. In the posterior part we find the third type of glands, the tubular variety.

The second division of the oviduct, as stated above, is the *albumin-secreting portion*. The funnel division gradually merges into the second portion. These two portions are distinguishable from each other. The walls of the albumin-secreting portion are much thicker and the longitudinal ridges are higher. This section is the longest of the five divisions, measuring from 40 to 42 centimeters in length, or more than half the length of the oviduct. The albumin division terminates rather abruptly into the third division, the isthmus (Fig. 59, *A*).

It is probable that the secretion of albumin is not confined to the cells of the second division; yet we are safe in saying that the major portion is formed here. The folds of mucous membrane in this division are thicker and higher than in the funnel, due to their containing cells of the high columnar type, and to the fuller development of the glands which are of the tubular variety. The muscular layer is heavier, and therefore the muscular power to force along the tubes' contents is greater. In the formation of the mucous folds we find, in transverse sections, that the central core is made up of connective tissue which carries blood-vessels and nerve filaments, as in other similar glandular structure. The epithelium contains glandular cells of two varieties, namely, the ciliated, columnar variety, and the unicellular, goblet variety (Fig. 36, No. *B*, 2). These two kinds of glandular cells are rather evenly distributed throughout the epithelium of this section of the oviduct. The unicellular gland cells are more numerous at the mouths of the ducts leading from the tubular glands. The nuclei of the ciliated cells are oval and lie near the middle of the cells or a trifle toward the base from the middle. The protoplasm of the cells is finely granular. Strong cilia in considerable number surmount each cell. In some cases the goblet or mucous cells have pushed apart the ciliated cells, and their prolongations extend farther than the surface of the ciliated cells. The nuclei of the goblet cells are round, and lie nearer the proximal end than those of the ciliated cells.

The third division of the oviduct, the *isthmus*, continues from the albumin-secreting portion and terminates in the expanded portion called by some anatomists, the uterus. Toward the posterior end of the albumin-secreting portion the longitudinal folds of mucous membrane become lower, making, at the juncture of this and the isthmus, a clear fine of demarcation. For a distance of 2 or 3 centimeters the folds are low, after which they gradually become higher, but never reach the height or thickness of those in the albumin-secreting portion.

The clear-cut line between the albumin-secreting portion and the isthmus is partly due to a zone in which the long tubular glands are lacking. The core of the folds of mucous membrane in this zone contain much more connective tissue. The cells are both ciliated and unicellular. The rest of the histological structure of the isthmus is the same as that of the albumin-secreting portion.

The function of the isthmus is to secrete, or to form, the shell membrane, the membrana testacea.

The fourth division of the oviduct is the *uterus*, or the shell-gland portion (Fig. 60, C). There is no clear line of demarcation between the isthmus and shell-gland portion, the walls gradually expanding. In this region the folds of mucous membrane become leaf-like and of considerable length, extending into the lumen, thus affording a greater cellular surface. The same coats of the duct are present here as in other parts; but the outer longitudinal muscular layer is thicker and possesses more strength. The quantity of connective tissue is about the same. The glands are of a tubular type and the same two varieties of epithelial cells are found here as elsewhere, namely, the ciliated high columnar and the unicellular mucous variety.

In the active glandular cells of the shell-forming region the nuclei are small, dark staining, and lie toward the center of the cell. The chromatin granules and nucleoli take on a comparatively deep basic stain, but they do not show the intense stain found in the albumin and the isthmus region. The cytoplasm of the uterine tubular glands does not present as heavy a granular appearance as that of the albumin portion. These cells are diffusely granular, the granules appearing of one size and taking the stain faintly. The function of this portion is to secrete, or form, the hard calcareous covering which has been described at the beginning of this section.

The fifth division of the oviduct is the *vagina*. There is located, at the juncture of the shell-gland portion with the vagina, a strong sphincter muscle. The vagina is that constricted portion of the oviduct extending from this muscle to the cloaca. The mucous membrane forms low narrow folds with secondary folds, which appear continuous with those of the shell-gland portion. The core of these folds is composed of connective tissue. The vagina in the hen of average size measures from 12 to 13 centimeters long (Surface). The inner or circular muscular layer is well developed; it is much thicker than in any other part of the oviduct. This extra development gives the power necessary to successfully expel the egg. The outer longitudinal layer is not so well developed; its bundles are scattered throughout the connective-tissue layers. The egg is caused to move along in the oviduct by a successive series of contractions of the circular muscular fibers posterior to it. There are no tubular glands in this portion, but a simple layer of high ciliated columnar epithelium, and some goblet, or mucous cells. The cells on the surface generally are long and slender; in the grooves between the mucous folds the cells are shortest, reaching their greatest length at the tops of the folds.

The function of the vagina is the secretion, or formation, of the outer shell cuticle commonly called the bloom, and also in certain breeds, as indicated above, the tint.

The cloaca furnishes a passage way from the vagina to the external world by way of the anus. The walls of the cloaca contain glands.

The Ligaments of the Oviduct.—It is held in position by two ligaments, one dorsal and one ventral. The *dorsal ligament* of the oviduct is formed by a double layer of peritoneum with a very small amount of connective tissue interposed. The peritoneum is also reflected over the oviduct. The ventral ligament of the oviduct is narrower than the dorsal, but is similarly constructed. Both ligaments are rather veil-like in appearance. During the first four or five months of the growth of the young female fowl the development of the oviduct and its ligaments is in proportion to that of the body. With the elongation that takes place about the time of functionation, as described above, the ligaments enlarge in proportion to the enlargements of the oviduct. The dorsal ligament maintains a line of attachment to the body wall from the caudal end of the body cavity to the fourth thoracic rib. The ventral ligament elongates only slightly during this developing period. It becomes thicker and stronger and early develops a muscular coat. It also grows in width except at the caudal end. At this point the ligament is simply a mass of muscular tissue of the smooth or involuntary type. These ligaments are fan-shaped.

The muscle fibers of the dorsal ligament of the laying hen, have their origin in a line near the medial side of the dorsal margin. At this point the bundles of fibers are quite thick, but are spread out thinly toward the margin of the oviduct. Frequent anastomoses are noted. The muscular fibers become continuous with the circular ones of the oviduct.

The *ventral ligament* of the oviduct of the laying hen is largely a muscular cord 3 to 6 centimeters in diameter. The caudal end is thicker, becoming gradually thinner toward the anterior portion. The bundles of muscular fibers extend toward the oviduct blending with the circular fibers of that viscus. The ligaments terminate anteriorly in such a manner that they aid in forming the serous ovarian pocket, which guides the yolk into the fimbriated portion, or funnel, of the oviduct. The walls of the ovarian pocket are formed by the left abdominal air-sac, a part of the intestine, and the mesentery. The dorsal portion is formed by the roof of the abdominal cavity, and the ventral portion is formed by the dorsal wall of the air-sac. The medial, the anterior, and the lateral limit of the pocket are formed by a fusion of the wall of the air-sac to the mesentery and to the body wall. Posteriorly, the wall consists of the transverse part of the small intestine and the caudal portion of the left cæcum with their attached mesentery.

THE DUCTLESS GLANDS

These glands do not possess excretory ducts. They furnish materials which are added to the blood or lymph as it passes through them. The material from each gland is known as an internal secretion, or hormone. Some of these secretions are powerful materials and influence profoundly the body nutrition. The ductless glands are usually given as follows: the spleen, the lymph glands, the pineal gland, the pituitary body, the thyroid gland, the thymus gland, the adrenal glands, and the parathyroids. The spleen, the pituitary, the pineal, and the lymph glands are described in other sections.

The Thyroid Gland (Fig. 21, No. 16). *Location.*—The thyroid gland lies on the ventral side of the carotis communis at a point where the carotis communis touches the jugular vein, which is about the point of origin of the vertebral artery.

Shape.—The thyroid gland is small, oval or somewhat roundish, and red or rose-colored.

Structure.—The thyroid gland has a fibrous capsule, which sends into the interior septa which divide it into acini. These acini are closed and contain a fluid. The thyroid is a ductless gland. Short arteries from the carotis enter this gland, and some large veins connect it with the jugular vein. The lymph vessels which lie along the neck are closely connected with it and receive twigs from the gland. The minute lymphatic capillary endings are found in its septa and in its capsule. The acini are lined with a single or a double row of cuboidal secreting cells. There are two kinds of cells, namely, secreting and resting cells. The actively secreting cells secrete colloid.

The **parathyroids** consist of two small bodies attached to the lower pole of the thyroid.

The Thymus Gland (Fig. 47, No. 10).—The thymus gland is an organ of fetal and early baby chickhood. It soon undergoes retrogressive changes into fat and connective tissue. It is of epithelial origin, being formed in fetal life from the entodermal cells of the dorsal end of the throat fissure.

Location.—The thymus gland lies anterior to the thyroid, the latter lying the deeper.

Shape.—The thyroid consists of two lobes, which are united by connective tissue, and appears as a loop-like acinous gland lying along the neck and near the region of the bronchi and the jugular veins with fibrous extensions toward the head.

Structure.—The gland lobes are divided into lobules, which consist of a cortical and a medullary portion. The cortex consists of nodules of compact lymphatic tissue similar to those found in the lymph glands. These occupy the chambers formed by the septa of connective tissue. In the medulla there are a number of spherical, or oval bodies composed of concentrically arranged epithelial cells. These are known as Hassall's corpuscles, and represent only the remains of the original glandular epithelium. They are characteristic of the thymus gland. The thymus appears to be a type of lymph organ. Lymph vessels are rare; a few blood-vessels on the upper side form capillary nets.

The Adrenal Gland (Fig. 54, F). *Location.*—The adrenal gland, often called the suprarenal capsule, lies just anterior to the front part of the anterior lobe of the kidney, adjacent to the testicles in the male and to the ovary in the female. It is loosely attached by connective tissue to the posterior aorta and to the vena cava.

Shape.—It is yellowish-brown or reddish-pink in color, small, and of irregular formation.

Structure.—The adrenal gland consists of a cortical and a medullary portion, although these two parts are not distinctly marked. The cortical portion has columns which extend deeply into the gland, and the medullary portion sends columns into the cortical portion. Therefore, the two substances, lying side by side, form a cord-like structure.

It is probable that the cortical portion is derived from the ingrowths of the peritoneum, and the medullary cords from the sympathetic ganglion.

The cells are cylindrical or polygonal in shape, with an eccentric substance between the columns. The cords, or columns, form between them, elongated channels which extend into the interior of the gland and end as blind or cæcal extremities. Large ganglionic nerve cells belonging to the sympathetic system occur near the surface of the gland. The blood-vessels are not well developed in the interior of the gland but are numerous and of good size in the outer parts. Lymph vessels are also present.

Function.—The adrenals are ductless glands. They secrete an internal secretion, or hormone, which influences the tonus of the blood-vessels. An extract from these glands is called adrenalin.

THE RESPIRATORY APPARATUS (Figs. 50 and 61, A)

Owen says: "Notwithstanding the extent and activities of the respiratory function in birds, the organs subservient thereto manifest more of a reptilian than of the mammalian type of formation."

By the action of the respiratory organs certain chemical and physical changes take place in the blood. The chief of these consists in absorption of oxygen from, and giving off carbon dioxide to, the atmospheric air, the former changes being necessary for the elaboration of the fluid, the latter for the elimination of a substance which, if retained, would prove injurious. The organs of respiration are invariably adapted to the wants of the animal and the medium in which it lives.

In the bird, which breathes through its nose, the organs of respiration are nostrils, nasal chambers, pharynx, superior larynx, trachea, inferior larynx, bronchi, bronchial tubes, lungs, and air-sacs.

The Nostrils and the Nasal Chambers.—The nostrils of the bird open externally by two small elliptical openings, which pierce the upper mandible. Within each nasal chamber (Fig. 26, *A*) are three turbinated laminæ, or *turbinated bones*. The inferior one is a simple fold adhering to the lower and anterior part of the nasal septum. The middle turbinated bone is the largest. It is of infundibular form, and adheres by its base to the septum and externally to the side wall of the nose. It is convoluted with two and a half turns. The superior bone, of bell shape, adheres superiorly to the frontal bone. The internal turbinated bone extends toward the orbit; the external terminates in a cul-de-sac behind the middle turbinal (Fig. 26, No. *A*, 1 and 2, and *G*).

The nostrils are separated by a partition which is partly bony and partly cartilaginous. The posterior nares is represented by a long slit in the hard palate.

The Pharynx and the Superior Larynx.—A transverse row of horny, filiform papillæ marks the anterior border of the pharynx, where in other animals the soft palate, or velum, is located. (For further description of the pharynx see special chapter.)

The supero-posterior border of the larynx, at the juncture of the larynx and the esophageal margin, is marked by a second transverse row of horny, filiform papillæ, which point backward. There is no epiglottis. The superior part of the larynx is pierced by an oval, slit-like opening, the *glottis*, which is provided with two lips. These when brought together, tightly close the glottis so that nothing can fall through into the larynx in the act of deglutition. The margin, or rim, of this opening is called the *rami glottis*. The glottis is controlled by two pair of muscles.

The superior surface of the larynx is somewhat triangular with the apex directed forward. A few delicate, filiform papillæ are upon its surface. The bird has, as already indicated, two larynxes, the *superior larynx* located at the upper end of the trachea, and the *inferior larynx* at the bifurcation of the trachea. The inner surface of the superior larynx is smooth and does not contain vocal cords; it is in these animals simply a passage for air. It is joined to the trachea inferiorly by a ligament, the *crico-trachealis*, and lies at the base of the tongue supported by two cornua of the os hyoideum.

The cartilages forming the principal support of the superior larynx, consist of four pieces, as follows: one unequal ventral piece, two side pieces, and one unequal dorsal piece. The cartilaginous, flat, ventral *cricoid*, early in the bird's life, often becomes bony. The side pieces are separated from it, only exceptionally fusing with it. The dorsal cricoid piece also often becomes bony. The two *arytenoid cartilages*, joined with the cricoid superiorly, are three-sided, and are united to each other in a sharp angle. They form the superior opening of the superior larynx.

The Trachea.—The trachea is cylindrical and varies in length in different kinds of birds in accordance with the length of the neck. It consists of from 90 to 120 cartilaginous rings, complete with the exception of the two uppermost, which rings are held together by intercartilaginous ligaments. The tracheal rings are constructed of hyaline cartilage and the ligaments of fibrous tissue. It is lined with a mucous membrane covered by columnar epithelium. The trachea is a passage for air alone and terminates in the inferior larynx.

The Inferior Larynx.—The inferior larynx, called the *true larynx* because it is the organ of voice, is located at the inferior end of the trachea and the superior ends of the bronchi. By some anatomists this organ has been called the *larynx broncho-trachealis*. The larynx is flattened laterally in fowls. It contains two membranous folds, which in the production of sound are caused to vibrate. These folds are half-moon shaped elastic structures, located in the bony, arrowlike way, intero-inferiorly. These structures are called the *membrana tympana interna*. In the duck this inferior larynx is represented by a drum-like cartilaginous and bony structure, called the *bulla tympaniformis*. This bulla is a resonant apparatus which serves to strengthen the voice.

In song birds there is a double glottis, usually produced by a bony bar, called the *pessulus*, or *os transversale*, which traverses the lower end of the trachea from front to rear. It supports a thin membrane which ascends into the tracheal area, and, terminating there by a free concave margin, is called the *membrana semi-lunaris*. This is most developed in singing birds, and being vibratile, forms an important part of their trilling vocal apparatus. The air passes on each side of the membrana semi-lunaris and its sustaining bone to and from the bronchi and lungs.

The last ring of the trachea usually expands as it descends, with its fore and posterior parts produced, and the lower lateral borders concave; the extremities of the pessulus, butts against the angle thus formed and expands to be attached, also with the fore and posterior terminations of the first half ring of the bronchus, strengthening and clamping together the upper part of the vocal framework. The second bronchial half ring is flattened and curved with the convexity outward, like the first, but is more movable. The third half ring is less curved and further separated from the second, to the extremities of which its own are connected by a ligament, and, for the intervening extent, by a membrane; its inner surface supports the fibrous cord, or fold, which forms the outer lip of the glottis of that side; it is capable of rotary movements on its axis, and is an important agent in the modulation of the voice. All these parts just described are bony.

The Bronchi and the Lungs.—The bronchi, two in number, are provided with only incomplete cartilaginous rings. They enter the inferior face of the lungs, toward their anterior and middle thirds and break up into primary bronchi, which give off at right angles, secondary bronchi, and these latter in turn give off tertiary branches.

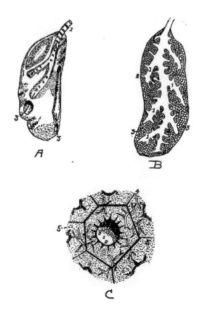

FIG. 61.—The lung.

A. The outer surface of one lung. Note the flattened oval shape. It is not divided into lobes. 1, The bronchus. 2, Primary tubules showing openings leading from the primary tubules to the secondary tubules. 3, Openings of two of the large tubes into the diaphragmatic and abdominal air-sacs.

B. Sectioned surface of lung. 1, Secondary tubules. 2, Tertiary tubules. 3, Interlacing capillaries and air cells.

C. 1, Cavity of tubule. 2, Its lining membrane supporting blood-vessels with large areolæ. 3, Perforations in the membrane at the orifices of the lobular passage. 4, Interlobular space containing the terminal branches of the pulmonary vessels supplying the capillary plexus, 5, to the meshes of which air gets access by the lobular passage.

The lungs occupy only about one-seventh of the thoracic space. They are long, flattened, and oval, extending along each side of the spine from the second dorsal vertebra to the anterior end of the kidneys, and laterally to the juncture of the vertebral with the sternal portions of the ribs. They present two faces, a *superior convex* and an *inferior concave*; two borders, an *external* and an *internal*; and two extremities, an *anterior* and a *posterior* (Fig. 61). The convex surface is also called the dorsal, costal, or superior face. It is moulded on the walls of the thorax and occupies a part of the intercostal space, pushing the intercostal muscles outward. When the surface of the lung is examined it is seen to be furrowed where the ribs pressed during life. These furrows are as deep as the ribs are thick. The sides of the lungs are covered with connective tissue which attaches them to the costal walls.

The concave, or inferior, face also called the *diaphragmatic* or *visceral face* is directed downward. The diaphragm separates it from the abdominal viscera. The surface is covered by connective tissue which closely attaches it to the diaphragm. It is perforated by the five tubules which bring the posterior air-sac into communication with the lungs.

The borders of the lungs extend parallel to the long axis of the body. The internal border is rectilinear, thick, and rounded. The external border is convex, thin, and sharp.

The anterior extremity terminates in a sharp point which occupies a space formed by the ribs externally and the inferior spines internally. The posterior extremity is somewhat rounded and extends as far back as the anterior border of the kidneys.

As soon as the bronchi enter the lungs they become broadened, the cartilaginous rings disappear, and they continue as membranous channels whose diameters gradually decrease, as they extend backward, to the point where they terminate in the *ostium caudale*, at which point they are surrounded by a cartilaginous ring. The ostium caudale brings the tubules into communication with the ventral air-sacs.

Twelve *air tubules* have their origin from each common bronchus, or trunk. Four are given off from the internal wall of the main bronchus by a series of openings arranged in a row. Seven are given off from the external wall by a second series similar to that of the first. The twelfth extends from the inferior wall, and immediately takes a course downward and outward and communicates with the posterior diaphragmatic air-sac. This may be considered as the terminal branch of the trunk.

All of these secondary canals, except the last, pass toward the periphery of the lung. They divide and subdivide at the periphery, covering it with their ramifications. The canals extending from the inner wall are distributed to the inferior face of the lung. Those extending from the outer wall are distributed to the outer face of the lung. The first constitute the *diaphragmatic* and the second the *costal bronchial tubes*.

The four *diaphragmatic bronchial tubes* are numbered in the order in which they are given off. The first is carried forward horizontally, the second transversely inward, the third obliquely inward and backward, and the fourth directly backward. They have, by some anatomists, been called the anterior, the internal, and the posterior diaphragmatic bronchial tubes. There are two posterior diaphragmatic bronchial tubes; the larger called the great posterior, and the smaller, which passes directly backward, the small posterior.

The *costal bronchial tubes*, seven in number, are numbered from the front backward in the order they are given off. Parallel at their origin, and side by side, like pipes of an organ, they soon spread out in fan-shape like the preceding. They extend from their central origin to the periphery. The first extends obliquely upward and inward to the anterior extremity of the lung. All branches from this bronchus extend from its anterior wall. The first branches are inflected to reach the external border of the lung. The succeeding branches are directed forward and the last forward and inward. They all meet those from the anterior diaphragmatic bronchus, but do not anastomose with them.

The second, the third, and the fourth costal bronchi extend in a transverse manner and ramify on the inner border of the lung.

The fifth and the sixth are directed toward the posterior extremity of the lung. The seventh, very small, reaches this extremity, where it disappears.

The first costal bronchus is the largest; those following it gradually become smaller. At their points of origin they adhere closely to the ribs. They are all imperforate, which is a distinguishing feature from those occupying the opposite face.

The *canaliculi* or *tertiary tubules* given off by these secondary bronchial tubes do not differ greatly in caliber in the various parts. They are given off at right angles from the pulmonary wall of each bronchus, and extend perpendicularly into the lung substance. Thus we find three kinds of conduits, the primary, the secondary, and the peripheral, or tertiary. The first are like the barbs of a feather on its shaft; and the second and parenchymatous are implanted on the pulmonary walls of the first, like the hairs of a brush on their common base. Thus instead of the branching of the bronchi being dichrotomous, as in mammals, it is piniform.

The canaliculi, or finer tubules, communicate with one another. The inner microscopic appearance of the canaliculi indicate that they are divided into areola, which gives them a cellular aspect. These tertiary bronchi open on a dense labyrinth of blood capillaries (Fig. 61, C). At this point the ciliated epithelial cells give way to simple squamous epithelium.

Thus we find three kinds of bronchi, or their ramifications, as follows: the primary, the secondary, and the tertiary.

The Air-sacs (Fig. 61, *A*).—The air-sacs are bladder-like structures consisting of a delicate cellulo-serous membrane, an extension from the bronchial tubes, in some places strengthened by an external envelope of elastic fibrous tissue. Long thin blood-vessels are distributed in the substance of these walls. They are branches from vessels of the general circulation and not extensions from those of the lungs. No lymphatics have been found in the air-sacs.

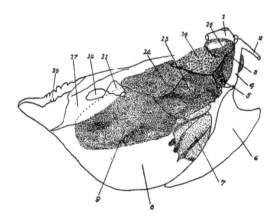

FIG. 61, *A*.—Diagram of air-sacs and their location. 1, The proximal end of the humerus. 2, The proximal end of the right clavicle. 3, The cervical air cell.

4, The right coracoid bone. 5, The anterior thoracic air cell. 6, The right side of the sternum. 7, The right side of the liver. 8, The peritoneum. 9, The right abdominal air cell. 10, The coccyx. 11, The proximal end of the right femur. 12, The right supero-posterior air-sac. 13, The right infero-posterior air-sac. 14, The right lung. 15, The axillary extension of the air-sac. 16, The obturator foramen. 17. The pelvis.

These sacs do not communicate with each other and normally they are not fully inflated. In some locations they extend into the bones and are in communication with the extensions of the bronchial tubes. In fact, by some anatomists they have been called "bladder-like, extra-pulmonary expansions of the bronchial tubes, free from cartilage." The air-sacs make the bird's body lighter, thus making long-continued flight possible. They are best developed in those birds which fly most. There are four pairs of cells and one single cell from which all other expansions and extensions are made. These sacs are as follows: a single anterior thoracic, and, in pairs, cervical, anterior diaphragmatic, posterior diaphragmatic, and abdominal.

The Anterior Thoracic Air-sac.—The anterior thoracic air-sac is located above the clavicles and the interclavicular space, in the cavity of the thorax. It is related superiorly with the trachea and the esophagus; laterally with the lungs and the cervical air-sacs; inferiorly with the sternum, the clavicle, and the interclavicular aponeurosis; posteriorly with the heart and the anterior diaphragmatic reservoir; and anteriorly with the integuments of the neck. It contains the inferior larynx and the two primary bronchi, and large vascular trunks from which are given off vessels supplying the neck and the wings.

Three prolongations, called subpectoral, subscapular, and middle, or humeral, arise from the lateral walls of this air-sac. These prolongations cross the walls of the thorax and pass around the articulation of the shoulder.

The *subpectoral prolongation* extends from the thoracic reservoir by an orifice situated behind the coracoid, and passes beneath the tendon of the great pectoral muscle. When the pectoralis major contracts, this contraction dilates the subjacent cell and draws into it a greater quantity of air.

The *subscapular* and the *humeral prolongations* communicate with the thoracic air cell by a common opening situated behind the small adductor muscle of the humerus. The subscapular prolongation, after leaving this point, spreads under the scapular and the subscapular muscle, which it separates from the ribs and corresponding intercostal muscles, and extends in a longitudinal direction.

The humeral prolongation, smaller than the subscapular, occupies the axilla, and is in shape triangular. It has from its summit into an infundibular fossa, an extension which enters the canal of the humerus. The walls of this cell form the lining of the air space in the humerus.

The thoracic air-sac thus possesses numerous membranous folds which divide its cavity. The contiguous structures which it overlies, as the trachea, the esophagus, the muscles of the inferior larynx, as well as the arteries and veins, make its outer walls irregular. This orifice is dilated during inspiration, by the contraction of the two first fasciculi of the diaphragm.

The Cervical Air-sacs.—The two cervical air-sacs are located just above the thoracic air-sac at the inferior part of the neck and in front of the lungs. They are cone-shaped with the base directed forward and the apex backward. They are related superiorly with the cervical muscles, and inferiorly with the thoracic air-sac from which they are separated by the trachea, the esophagus, the pneumogastric nerve, and the jugular veins. The walls touch each other internally, and form a median septum which includes in its substance the two common carotid arteries. Externally they are related to the origin of the cervical nerves, to each of which they contribute a small sheath. They surround the vertebral artery, and are connected with the subcutaneous muscles and the skin. The summits communicate with the anterior diaphragmatic bronchus. Prolongations extend from their bases which conduct the air into all the vertebræ of the neck and the back, into all the vertebral ribs, and into the spinal canal. Parallel with and adjacent to the vertebral arteries, and lodged in the canals excavated in the transverse processes of the cervical vertebræ, are two cervical prolongations, one on each side, which extend to the cranium from the base of the cervical reservoirs. From their sides, at the last six cervical vertebræ, are six extensions in the form of diverticuli, which, lying against each other, pass from each side into the muscles of the neck. They are surrounded by a thin fibrous envelope, a continuation of the mucous lining of the sac, and apparently form a canal in the inferior part of this region. These prolongations are better developed in palmipedes than in chickens. On the internal side of these prolongations, one or more foramina penetrate the vertebral segment, which allow the extensions of the prolongations into the spinal canal. Chauveau states that "as the medullary tissue is replaced by air in the bones of birds, so is the subarachnoid fluid replaced by air around the spinal cord."

The prolongations extending from the cervical air-sacs, having entered the thorax, terminate by passing into the first dorsal vertebra. After permeating every part of this vertebra, it escapes by a lateral opening and forms a small sac located between the first two ribs, near the origin of the first dorsal nerve. From this sac an extension is given off, which enters the second vertebral segment at the antero-lateral part; from this point it passes back, forming a new air-sac between the second and third ribs. It now passes in the same manner into the third vertebra and extends through the third intercostal sac, and so on till the last dorsal vertebra has been served. At the same time that these sacs receive the air from the vertebræ preceding them, and transmit it to those which follow, they communicate it to all the vertebral ribs. The aerial currents which leave the cervical air-sacs do not communicate with those of the cranium. Experiments show that the cranial bones have apparently no communication with the respiratory apparatus.

The Anterior Diaphragmatic Air-sac.—The two anterior diaphragmatic or supero-posterior air-sacs are related with the lungs anteriorly, and with the abdominal viscera posteriorly. Anteriorly also is the thoracic air-sac, posteriorly are the posterior diaphragmatic air-sacs, and laterally the ribs and the intercostal muscles and internally is the esophagus. The lungs communicate with these air-sacs through circular openings from the great posterior diaphragmatic bronchus and frequently by a second opening from this same tube. These are the only sacs which receive air from the lungs through two openings.

The Posterior Diaphragmatic Air-sac.—The two posterior diaphragmatic, or infero-posterior air-sacs are oval in shape and located between the thoracic and the abdominal cavity. They are related anteriorly with the anterior diaphragmatic air-sacs. These two sacs form a vertical transverse partition. The posterior diaphragmatic air-sacs are related posteriorly with the abdominal air-sacs from which they are separated by the diaphragm. They are related below with the lateral parts of the sternum and the sternal ribs, and externally with the ribs and the intercostal muscles. These air-sacs communicate with the lungs through openings located in the middle part of the external border of the lung, into the extremity of voluminous bronchial tubes which follow the direction of the largest air tubes.

The Abdominal Air-sacs.—The two abdominal air-sacs located on each side of the abdominal cavity, when inflated with air, form enormous bladder-like structures. They are related laterally with the abdominal wall and internally with the abdominal viscera. The anterior extremities are in communication with the mesobronchi and are somewhat inflected to pass under the fibrous arches extending from the spine to the pelvis. Anteriorly these sacs adjoin the diaphragm, the testes in the male, and ovary in the female, and to the parietes of the abdomen and those of the pelvis. Below and in front, they rest on a fibrous septum, which in all birds divides the abdominal cavity into two smaller cavities: one anterior, representing the abdomen and containing the liver; the other posterior, representing the pelvis and containing the gizzard and the intestines. The anterior portion overlies the posterior part of the lobes of the liver, the proventriculus, the spleen, and the gizzard. The kidneys are located above these air-sacs. Dorsal to the sacs is also a part of the intestines and in the female the oviduct. The abdominal air-sacs are attached by a ligament-like structure in their medial, their anterior, and their lateral margin. The posterior, the dorsal, and the ventral margin are free. Mesially this attachment is to the mesentery, connecting the left cæcum to the dorsal margin of the gizzard, and also to the mesentery of the proventriculus. The anterior attachment is to the body wall and extends in front of the end of the ovary and the adrenal glands. At the antero-lateral part of the body cavity the attachment extends in a widening band along the lateral side of the ovary and of the oviduct, as far back as the caudal margin of the sac. The lateral attachment is related to the kidney, the dorsal ligament of the oviduct, and the abdominal wall.

Each of these abdominal sacs has three extensions: one suprarenal and two femoral.

The *suprarenal extension* leaves the principal sac at the postero-external part of the kidney, extends upward, and forward, and expands over the surface of the kidney. At the internal border of the kidney, this prolongation extends between the transverse processes of the sacral vertebræ, reaches a height of the first dorsal vertebra, forms a triangular canal located above the sacrum in the sacral channel, and is separated from its fellow by a series of corresponding spinous processes.

The two *femoral extensions*, an anterior, small, and a posterior, large, extend from the abdominal air-sac at the cotyloid cavity, leave the pelvis through the bony passage occupied by the crural vessels, extend around the coxo-femoral articulation, and terminate in a blind extremity. In some birds, particularly in birds of prey and ostriches, there are prolongations extending into the femur, entering through a foramen at the anterior part of the great trochanter.

Summary of Bones Supplied by Each Air-sac.—The thoracic air-sac communicates on each side of the thorax with twelve bones, including the four sternal ribs. It supplies air to the clavicles, which are perforated at both their extremities, and to the coracoids, which are perforated just below their scapular extremity. The sternum is supplied through two series of openings, the middle ones that conduct air into the sternal ridge and the lateral ones, eight in number and very small, correspond to the intercostal spaces. The sternal ribs are penetrated by small foramina at their inferior extremities. From the subscapular extension the scapulæ receive air through one or two foramina at their anterior extremity. The humeral prolongation supplies the humerus through a foramen located at the upper edge of the humeral fossa, at the infero-internal part of the articular head.

The cervical air-sac furnishes air to all the cervical vertebræ, to all the dorsal vertebræ, and to all the vertebral ribs. The anterior parts of the vertebræ of the neck are supplied with air through the passage accommodating the vertebral artery. The posterior parts of the vertebræ are supplied by extensions from the interspinal canal. The first extensions obtain entrance to the anterior segments by one or more openings of the inner wall of the intertransverse canals; the median extensions penetrate the posterior segments by two openings, a right and a left, situated on the inner wall of those segments. The first dorsal vertebra is supplied with air in the same manner, by the middle and the lateral canals of the neck. This air, after passing through the first vertebra, leaves by a lateral exit to enter a small air-sac. From this it passes into the superior part of the second vertebra, escapes from this through its lower portion, to be received into a lateral sac, and so on to the last dorsal vertebra. These sacs also supply the vertebral ribs with air, which enters them by very small openings located on their spinal extremities.

The diaphragmatic air-sacs do not have communications with the bones.

The abdominal air-sacs communicate with the sacrum, the coccygeal vertebræ, the iliac bones, and the femurs. The air passing through the sacrum, the coccyx, and the ilium comes directly from the suprarenal extensions; the air which fills the femoral cavity comes from the femoral extensions.

In some birds these air spaces are more greatly developed than in others. The bones that are always aerated in all birds are the cervical and the dorsal vertebræ, the sternum, and the humeri. Those aerated in some kinds only are the furculum, the scapulæ, the vertebral and the sternal ribs, the sacrum, the coccyx, and the femurs. Those that are never aerated are the bones of the forearm, the hand, the leg, and the foot.

The service of air to the bones in most parts of the body by the air-sacs, as just shown, is in special cases otherwise rendered. The Eustachian tubes furnish air to the bones of the cranium and to the upper jaw; while the lower jaw receives air from the pneumatic foramen situated upon each ramus behind the tympanic articulation, and from an air cell which surrounds the joint.

The cavities of the embryonic bones, which afterward become pneumatic, are filled with marrow. Selenka states that the invasion of the bones by the air is a late development, and that in the humerus this invasion occurs after the twenty-second day in the life of the chick.

Hunter and Compar, who have made extensive researches, consider the function of the air-sacs as threefold.

First, the air-sacs are subsidiary respiratory organs, which aid in ridding the blood of waste products and in taking in oxygen.

Second, they aid mechanically the actions of respiration in birds. During the act of inspiration the sternum is depressed, the angle between the vertebral and the sternal ribs is made less acute, and the thoracic cavity proportionately enlarged; the air then rushes into the lungs and into the thoracic receptacles, while those of the abdomen become flaccid. When the sternum is raised, or approximated toward the spine, part of the air is expelled from the lungs and the thoracic air-sac through the trachea, and part is driven into the abdominal receptacles, which are thus alternately enlarged and diminished with the expansion and the contraction of the thorax. Hence the lungs, notwithstanding their fixed condition, are subject to due compression through the medium of the contiguous air receptacles, and are affected equally and regularly by every motion of the sternum and of the ribs.

Third, they reduce decidedly the specific gravity of the whole body. This must necessarily follow from the large spaces filled with air as well as from the absence in the bones of marrow and other fluids. The air-sacs by their position also render equilibrium more stable.

ANGIOLOGY

The Circulatory Apparatus.—The circulatory apparatus consists of two tubular systems: the blood vascular system and the lymphatic system.

The blood vascular system consists of the heart, the arteries, the veins, and the capillaries.

The heart is the central, propelling organ. The arteries form a series of efferent tubules, which, by branching, constantly increase in number and decrease in caliber, and which serve to carry the blood from the heart to the tissues. The capillaries are extensions from these latter tubules into which the arteries empty, and through the walls of which the interchange of elements between the blood and the other tissues takes place. The veins form a system of converging tubules which receive the blood from the capillaries, decrease in number and increase in size as they approach the heart, and return the blood to that organ.

The lymphatic system consists of capillaries and veins alone. As in the blood system, the lymph capillaries collect the effete material and pour it into the lymph veins, and these in turn, carry it to the large blood veins adjacent to the heart.

Both these systems have one and the same continuous lining, which consists of a single layer of endothelial cells. In the heart this lining is called the endocardium, and in the vessels, the endothelium. It forms a perfectly smooth surface.

THE HEART (Fig. 21, No. 7)

The heart of the domestic fowl is located in the median line of the thoracic cavity. It is more anterior and mesial than in mammals. Its axis is parallel with the axis of the trunk. The lungs being confined to the dorsal part of the trunk, the lower part of the heart is not surrounded by them, but extends backward, the apex resting in the anterior part of the anterior median fissure of the liver.

The heart has the form of an acute cone (Fig. 50, No. 1), the apex of which is bluntly rounded.

The heart is surrounded by a sero-fibrous sac, the *pericardium*. This sac adheres to the cervical air reservoirs anteriorly and to the diaphragmatic septum posteriorly. It is composed of two membranous layers: the *parietal*, external, dense, and fibrous; and the *visceral*, internal, and serous. The pericardial sac has no direct attachment to the heart, except at the upper extremity where it surrounds the large vessels emerging from it. The serous layer is reflected over the outer portion of the heart, where it is called the epicardium. The function of the pericardium is to prevent friction during the beating of the heart. It contains a small amount of serous fluid for perfect lubrication. This fluid is called the liquor pericardii.

Internally the heart has four cavities: two auricles and two ventricles. The *right ventricle* is more crescent-shaped than in solipedes, and in a manner envelops the left ventricle in front and to the right, though it does not reach the point of the heart. The *right auricle* is larger than the left. The *auriculo-ventricular* valve is not tricuspid as in mammals. This valve instead of being formed as usual by a membranous curtain, with margins retained by cords fixed to the walls of the ventricles, is composed of a wide muscular leaf, which appears to be a portion of the inner wall of the ventricle detached from the *interventricular septum*. This septum is convex; and the auriculo-ventricular orifice is an oblique slit situated between it and the muscular valve in question; so that, when the heart wall contracts at the systole, the valve is applied against this septum and closes the passage. The bicuspid, or auriculo-ventricular valve of the left side usually has two segments, though occasionally there may be three. The *fossa ovalis* is a depression behind the posterior semi-lunar valve in the septum of the heart. The membranous septum closing the foramen ovale is complete and strong but thin and transparent. The right auricle receives the blood from the two venæ cavæ coming from the anterior extremity, and from the posterior vena cava. These empty into a sinus. The left auricle has two vessels, the pulmonary veins which bring blood to it from the lungs.

Structure of the Heart.—The heart is lined by a serous membrane, the *endocardium*, which is a continuation of the endothelium of the blood-vessels. There are a few muscular pillars in the inner wall, called the *columnæ carnæ*. To give the heart its pumping power, it is made up of contractile tissue, a specialized kind of muscle called *heart muscle*. It is involuntary-striated and occupies an intermediate position, both morphologically and embryologically, between smooth involuntary muscle and striated voluntary muscle (Fig. 74, No. 4). It, like striated voluntary muscle, is both transversely and longitudinally striated. Heart muscle cells are short, thick cylinders, which are joined end to end to form long fibers. By means of lateral branches the cells of one fiber anastomoses with cells of adjacent fibers. Each cell of heart muscle contains one centrally located nucleus. There is no distinct sarcolemma, but the sarcoplasm is more dense near the surface of the cell, which gives it the appearance of an enveloping cell wall. There is a zone free from fibrillæ around the nucleus. The longitudinal fibrillæ, which make up the cell, are held together by a cement-like substance.

The main mass of the heart wall, called *myocardium*, consists of the specialized muscular tissue just described. The myocardium differs in thickness in different parts of the heart wall. It is thickest in the left ventricle and thinnest in the auricles. The left ventricle forces the blood through the systemic circulation and hence must be thicker to give it more power than is needed for the right ventricle, which forces the blood only through the lungs. The auricles are thinnest of all; for they receive the blood and pass it only to the chambers below. The *auricular appendages* at the base of the heart in fowls are not so well marked as in mammals. The auricular muscles consist of an outer coat common to both auricles, the fibers of which are transverse and of an inner coat, independent for each auricle, the fibers of which are longitudinal. Between the two coats, occur bundles of muscle the fibers of which run in various directions. The disposition of the muscle tissue of the ventricles is much more complicated. It is composed of several layers of fibers intricately interwoven.

The *endocardium*, covering the inner surface of the myocardium, forms a serous lining of all the chambers of the heart. At the arterial and venous openings it is continuous with and similar in structure to the intima of the vessels. The endocardium consists of two layers, an external layer closely attached to the myocardium and consisting of mixed fibers, including those of elastic tissue and smooth muscle cells; and an inner, single layer of endothelial cells, spoken of above.

The heart is supplied with nutrient blood by the two *coronary arteries*, which are given off from the common aorta just above the semi-lunar valves.

A right, or anterior, and a left, or posterior, coronary from their point of origin, turn in a ventral direction between the root of the aorta and the pulmonary artery, the right going to the right coronary groove and the left to the left coronary groove in the crown furrow. From here they send branches into the heart. The anterior, right, coronary, or coronaria dextra, the larger, is given off from the inferior wall of the aorta. It divides into a ramus superficialis and a ramus profundus. The ramus superficialis enters the crown furrow and divides into two or three branches on the right heart wall. These branches extend to the apex of the heart. Twigs from this artery along its course extend into the muscular wall reaching the posterior of the coronary groove where they anastomose with those of the left coronary, the ramus profundus, and with other branches from the same artery. The ramus profundus, larger than the preceding, gives off fine branches into the walls of the aorta and of the pulmonary artery, then enters from behind into the right wall of the septum ventriculorum, extends into the apex of the heart, and supplies the septum, or right inner chamber wall with the last branch, this breaking through the posterior wall of the auricular appendage.

The posterior, *left coronary*, or *coronaria sinistra*, originates from the dorsal wall of the aorta, proceeds as one branch on the upper surface of the left auricular appendix, and then extends between the left appendix and the pulmonary artery to the ventral surface of the heart. On the left side it supplies the wall of the pulmonary artery and gives off a ramus profundus. It sometimes divides into two parts and supplies the ventral wall of the right chamber and then extends to the left wall of the septum medium. The rest of the coronaria sinistra enters into the left crown furrow as the ramus superficialis, which provides the left and dorsal upper surface of the left chamber to the apex. From this furrow it extends into the left chamber and the left appendage, and finally fuses with the ramus superficialis of the coronaria dextra.

THE BLOOD-VESSELS

The blood-vessels consist of arteries, veins and capillaries.

The Structure of the Capillaries and Arteries.—The capillaries are minute vessels which connect the *arterioles*, or terminal arteries, with the *venules*, or terminal veins. They are only from 6 to 14 microns in diameter. Their walls consist of a single layer of endothelial cells, which are somewhat elongated in the long axis of the vessels. Their edges are serrated, and are united by a small amount of intercellular cement-like substance. Capillaries branch without diminution in caliber, and these branches anastomose to form capillary networks, the meshes of which differ in size and shape in different tissues and organs. The largest meshed networks occur in the serous membranes and in the muscles; and the smallest occur in the glands, such as the liver.

The *walls of the arteries* are thick and stand open when empty, owing to the elastic tissue contained in their walls, while the walls of the veins collapse when empty, owing to their containing a smaller amount of elastic tissue. The arterial wall is provided with three coats: *tunica intima*, or inner coat; *tunica media*, or median coat; and *tunica adventitia*, or outer coat.

The *tunica intima* consists of a single layer of endothelial cells, continuous with and similar to that forming the walls of the capillaries. In passing from the capillaries to the arterioles, there is first a thin coat, or sheath-like layer, of connective tissue around the outside of the endothelial tubes. Further along, isolated smooth muscle cells arranged in a circular manner occur between the endothelial layer and the layer of connective tissue, this structure forming vessels called precapillary arteries. Further along still, the muscle cells form a complete layer; in this section the vessels are called arterioles and are made up of three coats: the inner endothelial, the middle muscular, and the outer fibrous.

In arteries of medium size the intima consists of the endothelial layer, a layer of delicate white and elastic fibers, connective-tissue cells, and the membrana elastica interna, or an outer layer, the elastic layer, of the intima.

The *media* consists of a thick coat of circularly arranged smooth muscle cells, its thickness depending largely upon the size of the vessels. There is also a small amount of fibrillary connective tissue, which supports the muscle cells. Elastic tissue is present in the media, the amount depending on the size of the vessel, the larger the vessel the more elastic tissue there is present. In the large arteries coarse elastic fibers intermingle with the finer ones. When much elastic tissue is present the muscle cells are separated into more or less well-defined groups.

The *adventitia* is composed of loose connective tissue with some elastic fibers. A few smooth muscle cells are present, which, as are also the elastic fibers, are arranged longitudinally. The adventitia, blending with the connective tissue surrounding the arteries, serves to anchor the vessels to the surrounding structure.

Structure of the Veins.—In many respects the walls of the veins resemble those of the arteries. The same three coats exist and the same elements enter into their structure. The transition from capillary to small veins, and from those to larger veins, is similar to the transition from the arteries to capillaries, in inverse order. The walls are not so thick as those of arteries. The elastic tissue is much less in quantity and in smaller veins disappears. There is not a clear line of demarcation between the intima and the media.

The veins of birds differ from those of mammals in that they have fewer *valves*. The valves are also less perfect, and often permit a backward flow of blood.

The walls of the arteries and of the veins are supplied with nutrient blood-vessels. These are called the *vasa vasorum*, or blood-vessels of the blood-vessel wall. They are mostly in the adventitia. They may arise from the vessel to which they are distributed or take origin from an adjacent vessel. These small arteries supplying the vessel coat after terminating into capillaries form small veins through which the blood, from the structure of the vessel wall, is returned.

The walls of the blood-vessels are supplied with both medullated and non-medullated nerves. The non-medullated nerve fibers are axones of the sympathetic neurones and control the caliber of the vessels. These fibers are called the *vasomotor nerves*. They form plexuses in the adventitia, from which are given off branches which penetrate the media and terminate on the muscle cells. The medullated nerves are the axones of the spinal nerves. The larger fibers are found in the connective tissue outside the adventitia and give off branches to the media where they divide repeatedly, lose their sheath, and terminate in the media and at times in the intima.

THE ARTERIAL TRUNKS

The *common aorta* is short; it originates from the left ventricle (Fig. 61B, No. K, 14), and is guarded by three *semi-lunar valves*. The aorta breaks through the pericardium just to the right of the pulmonary arteries in a ventral direction; it then turns upward dorsally and to the right of the inferior bodies of the vertebræ. It is then directed anteriorly, and dorsally to the right bronchus, between the right bronchus and the right lung. The right and the left coronaries are given off from the common aorta; they have been discussed. There is next given off the left brachio-cephalic or *brachio-cephalic sinister artery* (Fig. 61B, No. K, 10), which is just above the border of the base of the heart. This artery passes upward and slightly forward, over the center of the inferior larynx. Just beyond this point the *subclavian artery* is given off; this artery later becomes the axillary and the axillary the brachial artery. Then there is given off the anterior and the posterior thoracic arteries; and finally, the pectoral, which supply the pectoralis muscles and later terminate in the carotid, the vertebral, and the cervical arteries. The carotid artery gives off an esophageal artery. The other arterial trunk given off from the common aorta is the right brachio-cephalic or brachio-cephalic dexter (Fig. 61B, No. K, 18). The right brachio-cephalic artery gives off the subclavian, which continues as the axillary, and continues as the brachial artery. There is given off the anterior and the posterior thoracic, the right carotid (Fig. 61B, No. K, 11), the vertebral, and the dorsal. The last continues as the cervical.

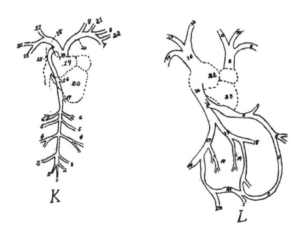

FIG. 61B.

K. The arterial trunks. 1, The middle sacral artery. 2, The hypogastric artery. 3, The posterior mesenteric artery. 4, The ischiadic artery. 5, The femoral artery. 6, The renal artery. 7, The left pectoral artery or thoracico caudalis. 8, The left axillary artery. 9, The left carotid artery. 10, The left brachio-cephalic artery. 11, The right carotid artery. 12, The right brachial artery. 13, The right pectoral artery. 14, The common carotid artery. 15, The posterior aorta. 16, The celiac axis. 17, The anterior mesenteric artery.

18, The right brachio-cephalic artery. 19, The auricular portion of the heart. 20, The ventricular portion of same. 21, Sterno-clavicular artery. 22, Anterior
thoracic.

L. The venous trunks. 1, The caudal vein. 2, The coccygeo-mesenteric vein. 3, The posterior mesenteric vein. 4, The anterior mesenteric vein. 5, The gastro-duodenal vein. 6, The portal vein. 7, The hepatic vein. 8, The left anterior vena cava. 9, The left pectoral vein. 10, The left brachial vein. 11, The left jugular vein. 12, The right jugular vein. 13, The right brachial vein. 14, The right pectoral vein. 15, The right anterior vena cava. 16, The posterior vena cava. 17, The common iliac vein. 18, The femoral vein. 19, The renal vein. 20, The internal iliac vein. 21, The hypogastric vein. 22, The auricular portion of the heart. 23, The ventricular portion of same.

The *pulmonary arterial trunk* is given off from the conus arteriosus of the right ventricle. It is guarded at its origin by three semi-lunar valves similar to those of the aorta. The trunk divides into two pulmonary arteries, which are short, one called the pulmonalis dexter and the other the pulmonalis sinister. The former goes to the right and the latter to the left lung. Each branch penetrates the lung near the bronchus. These arteries divide, or branch, similarly to the bronchi. Following the branchings of the bronchi, they finally terminate into the lung capillaries, forming networks on the bronchi and the air-tube terminals.

BRANCHES OF THE BRACHIO-CEPHALIC ARTERY

(Thyroidea

((Cervicalis inferior
(Vertebralis (Vertebralis anterior
((Vertebralis posterior
((Arteria cervicalis ascendens
(Bronchiales
(Truncus   (Inferior esophageal
(caroticus (Subcutaneous colli

(((Carotis cerebralis

((Carotis   (Cervicalis superior
((communis   (Occipitalis
(((Carotis externa

((Spinalis anterior (Basilaris (Cerebelli inferior

Brachio-cephalic (

((Clavicularis
((Sterno-clavicularis (Sternalis

(((Acromialis

((

(((Subscapularis

((((Ulnaris

(Subclavia (Axillaris   (Brachialis (Brachialis profunda

(((Radialis

((Humeralis   (Circumflex humeralis

(((anterior

(

((Thoracica externa

(Thoracalis   ((Arteria thoracica
(Thoracica inferior (longa

(

(Mammaria interna

(Circumflex humeralis posterior

Brachialis profunda (Collateral ulnaris

(Collateral radialis

The cord-like remnant of the *ductus botalli*, or embryonal connection between the lung arterial trunk and the anterior aorta, has been observed but is rare in grown birds.

The arteries, the veins, and the lymphatic vessels of birds anastomose far more frequently than those of mammals.

BRANCHES OF THE ARTERIA CAROTIS CEREBRALIS INTERNA

(Occipitalis   (Occipitalis sublimis
((Occipitalis profunda   (Meningea

((Temporalis
(((Rete temporale
(Ophthalmica   (Recurrent ophthalmicum (Rete ethmoidalis
(externa   (Ethmoidalis

((Plexus palpebralis

((Plexus alveolaris inferior

((Plexus muscularis
Carotis cerebralis (Plexus temporalis   (Plexus lacrimalis

((Ramus ciliaris posticus

((Meningea media
(
((Sphenoidea   (Ethmodalis

((Spheno-maxillaris   (Ethmoidalis (externa

(((Ophthalmica ((Ethmoidalis

(((interna   ((interna
(Cerebralis   ((
(Ramus   (Sylviæ

(anterior (Cerebralis profunda (Choroid plexus
(Arteria centralis retinæ
(Ramus posterior

BRANCHES OF THE ARTERIA CAROTIS EXTERNA

(Hyoidea

Carotis externa (Laryngea superior
(Facialis

(Lingualis

BRANCHES OF THE ARTERIA CAROTIS FACIALIS

(Auricularis

Facialis (Facialis interna (Alveolaris inferior (Mentalis

(Maxillaris interna

(Facialis externa

BRANCHES OF THE CAROTID TRUNK

The **carotis communis** artery springs from the carotid trunk of the brachiocephalic. It is directed horizontally, ascending to the ventral side of the neck. It then extends downward to the inferior median neck region. Just after leaving its origin it gives off several small branches to the bronchi and to the esophagus, and extends toward the head. The carotid lies on the thyroid gland and at this point touches the jugular vein (Fig. 21). At this point the **thyroid arteries** are given off from the carotid trunk. The thyroid gland also receives blood from the bronchialis artery.

Dorsalward and near the thyroid gland the carotid artery gives off a branch (the bronchialis) which accompanies the recurrent laryngeal nerve, along the inferior larynx and the bronchi, and supplies the lung substance and that part of the esophagus in this region.

The **vertebral artery** is given off from the carotid trunk, dorsalward to the thyroid gland, and on the left side. The right vertebral artery may be given off from the brachialis dextra or right brachialis.

The **inferior esophageal artery** is given off from the ventral side of the carotid, supplies the esophagus, extends then to the skin of the neck and to the trachea, is directed anteriorly toward the head and anastomoses with the vertebral artery. This artery accompanies the vagus nerve and forms a collateral artery to the carotid and the vertebral artery.

The **subcutaneous colli** springs from the carotid artery near the thyroid gland and communicates with the inferior cervical artery, which, in turn springs from the vertebral artery.

The Common Carotid.—Near the last cervical vertebra the two carotid arteries occupy the same channel, to which they are attached by fascia, and, lying close to the inferior surface of the bodies of the cervical vertebræ, are covered by the colli muscles. Near the third or the fourth cervical vertebra, the two carotid arteries separate. Near the atlas each carotid divides into the *carotis cerebralis* and *carotis externa*. Near this division the *superior cervical artery* branches off. This latter artery extends down the neck in company with the pneumogastric nerve and the jugular vein, supplies the skin and the neck muscles, and anastomoses with the inferior cervical artery and the subcutaneous colli artery.

The *occipital artery* originates either from the carotis communis or sometimes from the superior cervical.

BRANCHES OF THE CAROTIS CEREBRALIS

The branches of the carotis cerebralis are as follows:

1. The **occipitalis** (Fig. 72, No. 18), which in turn gives off, first, the occipitalis sublimis. The *occipitalis sublimis* supplies the outer and the middle portion of the digastricus and the posterior mylo-hyoideus muscle. It also gives off, second, the *occipitalis profunda* which supplies the inner portion of the digastricus muscle and becomes the *meningeal artery*, which passes through the foramen vagi, entering the brain cavity, where it supplies the coverings of the brain as far as the sella turcica. Superficial branches of this artery are distributed to the muscles in the region of the atlas and anastomose with the vertebral artery.

2. The **ophthalmica externa**, which passes below the articulation of the quadrate bone, around the tympanic cavity, enters the canalis caroticus, and passes into the cranial cavity. The ramus occipitalis, passing out of the same foramen, again enters the diploë of the cranium; then after passing dorsally over the upper outer semicircular canal of the ear, passes backward through the occipital bone.

The ophthalmica externa (Fig. 72, No. 9) forms a main trunk and gives off two branches: the *temporal artery* and the *recurrent ophthalmic*. The recurrent ophthalmic gives off the *rete temporalis*, *rete ethmoidalis*, a branch to the orbital gland, and finally anastomoses with its own branches and with those of the internal ophthalmic at the olfactory foramen and aids in forming the *ethmoidal artery*. These arteries supply blood by giving off branches to the muscles of the eye, sclerotic coat, iris, choroid coat, and the ciliary bodies. It gives off another branch to Harder's gland, and finally anastomoses with the ethmoidalis.

3. The **plexus temporalis**, rete mirabile ophthalmicum, or wonderful network (Fig. 72, No. 8), is formed between the second and the third trunk of the trigeminus nerve and the rete ophthalmicum.

Inferior to this is the *alveolar plexus*, which plexus accompanies the third branch of the trigeminus into the lower jaw. Two main branches form this plexus, one coming from the carotis facialis, which anastomoses with the alveolar artery.

The *palpebral plexus* lies between the trunks of the trigeminus, and supplies mainly the lower eyelid.

The *plexus muscularis* extends to the fifth portion of the temporal muscle.

The *plexus lacrimalis* forms on the posterior orbital wall on the ramus ethmoidalis and supplies the lacrimal gland and the upper eyelid. It anastomoses with branches from the facial artery.

The *ramus ciliaris posticus* supplies the inferior rectus and the external rectus muscle of the eyeball. It anastomoses with the ophthalmica externa.

The *median meningeal artery* passes through the foramen occupied by the second branch of the trigeminal nerve. Passing into the cranial cavity, it supplies the dura mater. Before passing into this foramen, small branches are given off, which supply the skin and the temporal muscle; and some branches anastomose with the superior cervical and the ethmoidal arteries.

4. **The Cerebral Artery.**—Each cerebral artery enters the cranial cavity through the canalis caroticus located in the sphenoid bone, passes forward medially from the cochlea, dorsally from the Eustachian tube, and passes through a small canal which opens on the inner surface of the sella turcica. Originating in this foramen, the *sphenoid artery* divides into two branches which anastomose with the pterygoidean and the pterygo-pharyngeal artery and which supply the upper jaw and the throat regions.

The *spheno-maxillaris artery* supplies the gums.

The right and the left cerebral arteries unite at the sella turcica. After this union they divide again immediately and pass to the base of the brain, where they give off twigs to the optic nerve, to the optic chiasm, and where also is given off the arteria retinæ centralis. Passing posteriorly each cerebral artery gives off the *ramus posterior*, and then passes to the side of the cerebellum.

The *basilar artery*, a continuation of the anterior spinal artery, is located ventrally and mesially to the cerebellum. Laterally and inferiorly the basilar artery gives off the inferior cerebellar artery.

The *ramus anterior* is given off from the cerebralis artery and continues as the *internal ophthalmic artery*. The ramus anterior also gives off the *sylvian artery*, which supplies the sides of the cerebrum and the middle brain. In the fissure between the hemispheres and the optic thalamus is located the *arteria cerebri profunda*. This artery passes along the median surface of the cerebrum and in its course supplies the adjacent parts. It enters into the formation of the choroid plexus of the lateral ventricle.

The *internal ophthalmic artery* extends out of the cranial cavity, through the optic foramen, into the orbital cavity. It passes upward along the interorbital wall and supplies the optic nerve, the trigeminus nerve trunk, and the eye muscles, and dorsally anastomoses near the olfactory nerve with the rete ethmoidale and the external ophthalmic artery, and continues as the ethmoidal artery.

The *ethmoidal artery* supplies the supra-orbital gland and the gland of Harder. It gives off twigs to the rete ophthalmicum and other branches to the skin of the frontal region, where it anastomoses with branches of the external facial artery. These branches in the frontal region are rather large. Large branches are given off to the comb.

The *arteria ethmoidalis externa* originates in the nasal cavity from a division of the ethmoidal artery.

The external ethmoidal artery passes below the lacrimal bone, extends forward, supplies the walls of the nasal cavity, and finally sends a branch anteriorly into the upper median portion of the jaw bone, and other branches to the septum nasi and other parts of the nasal cavity.

There are frequent anastomoses between the ethmoidalis interna, the facialis, and the spheno-maxillary artery.

The *internal ethmoidal* artery supplies principally the posterior turbinated bones and septum nasi.

BRANCHES OF THE EXTERNAL CAROTID ARTERY

Branches from the external carotid and the facial artery supply the tongue and its muscles, the larynx, the lower jaw bone, the gums, and the upper lateral facial region.

The external carotid (Fig. 62, No. *A*, 8) gives off the following branches:

First, the **hyoid artery** (Fig. 72, No. 12) which supplies the inner portion of the depressor mandibular and also the cornua of the os hyoideum, and extends to the tip of the tongue.

Second, the **superior laryngeal artery** (Fig. 72, No. 16) which supplies the sterno-brachialis and gives off branches which extend downward to the trachea and to the esophagus. The main artery passes along the left side of the esophagus and anastomoses with branches of the inferior esophageal, forming collateral circulation in that region.

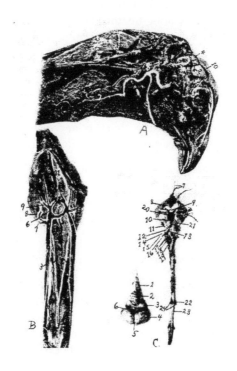

FIG. 62.

A. Blood-vessels of the head. 1, The cerebellum. 2. The cerebrum. 3, The semicircular canals. 4, Sinus occipitalis superior. 5, Sinus temporalis sphenoideus. 6, Sinus transversus dexter. 7, Superior esophageal artery. 8, Carotis externus. 9, Lingual artery. 10, Sinus longitudinalis.

B. Veins of the antero-inferior part of the head and neck. 1, The right carotid artery. 2, The left carotid artery. 3, Vena jugularis dexter. 4, Vena vertebralis dexter. 5, Transverse vein. 6, Vena cephalica dexter anterior. 7, Vena lingualis. 8, Vena cephalica posterior. 9, Vena facialis externa dexter. 10, Vena infrapalatina. 11, Posterior nares. 12, Vena facialis interna. 13,

Vena facialis externa sinister.

C. The brain. 1, Medulla oblongata. 2, Cerebellum. 3, Optic lobe. 4, Cerebrum. 5, Longitudinal fissure. 6, Transverse fissure. 7, Olfactory nerve. 8, Optic nerve. 9, Optic commissure. 10, Motor oculi. 11, Patheticus. 12, Trigeminal, or trifacialis. 13, Abducens. 14, Facialis. 15, Auditory, or acousticus. 16, Glosso-pharyngeus. 17, Pneumogastric. 18, Spinal accessory. 19, Hypoglossus. 20, Hypophysis. 21, Crus cerebri. 22, Ganglion on superior nerve trunk. 23, Spinal cord. 24, A pair of spinal nerves.

Third, the **lingualis artery** (Fig. 62, No. *A*, 9) which lies between the mylohyoideus and the posterior part of the hyoid bone. It supplies the tongue muscles. It passes to the median surface of the lower jaw bone, where are given off small branches which enter the jaw and anastomose with the inferior alveolar artery.

Fourth, the **Facial Artery**.—The facial artery divides into the following branches:

The *auricular artery* is given off near the articulation of the os quadratum with the os zygomaticum. It is located at the outer auditory canal, and its branches are distributed to the parotid region and to the depressor mandibular, or digastricus, muscle.

The *external facial artery* is located between the os quadratum and the masseter muscle. It gives off a branch to the lower jaw bone and to the skin of that region, and then passes to the lacrimal bone and supplies branches to the commissure of the mouth, the auditory canal, the masseter muscle, the three eyelids (upper and lower eyelids and the membrana nictitans), the nose cavity, and the skin in the frontal region. It communicates on the other side with the ethmoidal artery.

The *internal facial artery* passes over and through the os pterygoideum, supplies all the jaw muscles, and is continued as the inferior alveolar artery.

The *inferior alveolar artery* enters the canal of the lower jaw bone along with the mandibular nerve. It is finally continued as the mental artery, leaving the canal, and passing to the outer jaw surface.

The facialis continues as the *internal maxillary artery*.

The internal maxillary artery (Fig. 72, No. 15) supplies the pterygoid muscle, the upper part of the pharynx, the commissure of the mouth, the salivary gland region, the gums, and finally anastomoses with branches of the sphenomaxillaris artery.

Branches of the Vertebral Artery

The vertebral artery is given off, dorsally near the thyroid gland, from the carotis communis or from the carotid trunk. The vertebral artery passes horizontally and dorsally into the canals formed in the transverse processes of the cervical vertebræ. After it leaves the carotid, it is divided into anterior and posterior branches.

The **posterior vertebral artery** passes in the canals of the transverse processes of the fifth or sixth dorsal vertebra. It sends branches into the vertebræ, into the spinal canal, and into the intercostal muscles between the ribs and anastomoses with the intercostal arteries.

The **anterior vertebral artery** is larger than the posterior. It extends laterally along the side of the neck in the foramen of the transverse processes of the cervical vertebræ, and lies along the course of the inferior vertebral vein and the deep imbedded trunk of the sympathetic nerve. It continues to the head. In its passage it gives off to each vertebral segment a dorsal and a ventral branch. Twigs from these branches pass into the bodies of the vertebræ and the spinal canal, giving nutriment to the bony structure and to the spinal cord and its coverings. Other twigs are distributed to the muscles of the neck and some finally anastomose with branches of the carotis communis. This artery, reaching the head, gives off a long anastomosing branch which passes between the atlas and posterior part of the occipital bone and joins the ramus profundus and the occipital artery, thus again communicating with the carotid artery. The remainder of the vertebral artery is small and passes through the foramen magnum into the cranial cavity where it anastomoses with terminal branches of the cerebral artery. At the base of the neck and before the vertebral artery enters the canal of the cervical vertebræ, it gives off the *arteria cervicalis ascendens*, superior artery of the crop (Fig. 73, No. 18) which branches out on the upper surface of the crop is also distributed to the neck muscles, and later subdivides into the transverse cervical arteries which supply the skin and the muscles of the base of the neck, and the shoulder region.

On the ventral side of the neck there are given off two subvertebral carotid arterial branches. They lie in a shallow furrow on the ventral side of the cervical vertebræ and close to the median line.

Branches of the Subclavian Artery

The subclavian artery (Fig. 20, No. 13; Fig. 73, No. 11) gives off the following branches:

1. **Sterno-clavicularis** (Fig. 73, No. 9) which originates on the upper part of the subclavia, between the carotid artery and external thoracic (Fig. 73, No. 3), divides into many branches. The sterno-clavicularis gives off the *sternal*

artery, which enters by the side of the supra-coracoid muscle and the anterior rim of the sternum. It is distributed to the inner surface and over the air-sac. Another outside branch passes the posterior end of the crista sterna and supplies the large breast muscles.

The *clavicular artery* (Fig. 73, No. 8) accompanies the clavicle to the shoulder-joint.

The *acromial artery* is given off from the sterno-clavicularis on the ligament near the shoulder-joint.

2. The **thoracic artery** branches from the subclavian artery (Fig. 70, No. 4). The thoracic artery gives off the following branches:

The *internal mammary*, or *internal thoracic artery* (Fig. 73, No. 16) arises on the inner side of the sternum and extends downward and backward, giving nutrient branches to the anterior vena cava and to the diaphragm. At the point of the costo-sternal muscle it divides into an inner and an outer branch.

The inner branch supplies the costo-sternal muscle and extends along the ribs and along their juncture with the breast-bone. It finally extends posteriorly, giving numerous branches to the abdominal muscles. The outer branch is distributed in a similar manner, giving off branches to the diaphragm and to the abdominal muscles, and anastomoses, on the surface of the abdominal muscles, with the epigastric artery.

The *external thoracic artery* (Fig. 73, No. 3), ramus superior, supplies principally the large breast muscles.

The *inferior thoracic artery*, external ramus inferior, passes along the outside of the pectoralis major muscle, extends downward, and gives off the arteria thoracica longa (subcutaneous thoracic). This latter artery supplies the skin of the breast region. Other branches are given off to the muscles of the breast and to the skin of the region, and some finally anastomose with the branches of the sternal artery.

3. The **axillary artery** (Fig. 73, No. 12) extends out of the thoracic cavity along with the brachial nerve plexus. It gives off the *subscapularis* which supplies the muscles of the scapular region. The axillary artery terminates as the *brachialis* which passes between the biceps brachii and anconeus muscles downward along the humeral shaft. It gives off the anterior humeral circumflex artery and the brachialis profunda artery (Fig. 68, No. 3). It gives off near the elbow-joint the ulnar and radial arteries.

The *anterior circumflex humeral artery* passes through the short head of the biceps and gives off a branch to the biceps muscle, supplying the insertional part of the breast muscles.

BRANCHES OF THE BRACHIALIS PROFUNDA

The brachialis profunda artery (Fig. 68, No. 3) continues as the *posterior circumflex humeral artery*. This artery supplies the muscles of the posterior part of the humerus, the skin of the wing, and the muscles of the brachial region. The trunk of the arteria brachii profunda passes downward along the anconeus muscle, supplying that muscle and giving off the collateral ulnar artery. On the ulnar olecranon it anastomoses with the recurrent ulnar, thus establishing, at this region, collateral circulation. The rest of the brachialis sends branches to the biceps muscle and to the skin of the upper arm, and finally gives off the arteria collateralis radialis, which supplies the condyloulnaris. It then anastomoses with the end branches of the recurrent radialis artery.

The *radial artery* (Fig. 67, No. 4) passes downward on the outer rim of the forearm. It continues down to the carpal region, where it supplies the muscles and the skin of the carpal region it gives off the recurrent radial, which passes on the middle finger and finally anastomoses with the collateral radialis artery.

The *ulnar artery* (Fig. 68, No. 6 and 9) passes downward on the inner surface of the ulna to the carpal region where it gives off a branch to the wing plexus and divides into two branches, the smaller branch supplying the thumb (Fig. 68, No. 7). This branch passes down the radial side of the middle finger to the last joint of the second finger. The larger (Fig. 68, No. 8) lies between the second and the third finger bones, and passes through a slit between these two bones to the flexor side of the hand and extends to the last finger-joint, supplying the structures in the region. The ulnar artery gives off a small arterial twig to the papilla of each large wing feather (Fig. 67, No. 6).

The *recurrent ulnar* passes backward to the olecranon of the ulna and supplies the muscles, the skin, the feathers, and other structures of the region, and terminates in end collateral ulnar branches (Fig. 67, No. 5).

BRANCHES OF THE POSTERIOR AORTA[7]

(Esophageal

(Intercostales

(Dorsal

(Lumbars

(Spermatics
(Renals

(Ovarian
(Celiac   (Recurrent esophageal
(axis   (Recurrent intestinalis (Recurrent ilio-colicus
((Posterior or recurrent sinister   (Renalis
((Anterior or recurrent dexter   (Splenics

(((Hepatic
((Hepatica dextra (Gastric

(
Posterior (Sacralis media (Coccygeæ laterales
aorta   ((Coccygea media (Coccygeæ laterales
(
(Anterior mesenteric (Recurrent ilio-celiacus

((Recurrent superior hemorrhoidal

(Posterior mesenteric (Median hemorrhoidal
((Recurrent renalis
((Tibialis postica

(Ischiadica (Tibialis antica (Peroneal (Anterior tibial plexus

((Arteria ovarialis

(Pudenda communis (Renalis
((Hemorrhoidalis intima
((Pudenda externa
(
(Crural   (Internal pelvic (umbilical)
((External iliac) (Circumflex femoris
((Femoralis

<u>7</u>. Nomenclature used by Bronn.

The posterior aorta (Fig. 63, No. 2) passes backward along the inferior part of the bodies of the dorsal and lumbo-sacral vertebræ. In the thoracic cavity it lies dorsal to the esophagus. It gives branches (*esophageal*) only to the esophagus during its passage as far as the seventh dorsal vertebra. Following this point there are given of several pairs of arteries which divide into a superior and an inferior branch. The superior branch extends upward and anastomoses with branches of the vertebral artery and other twigs coming from the *intercostal arteries*. These latter arteries given off are small dorsal arteries. The inferior branches extend downward between the ribs and supply the intercostal muscles. The intercostal arteries do not take their origin from the aorta in numerous and regular branches as in mammals; they consist originally of but few vessels, which are multiplied by anastomoses with each other and with the arteries which come out of the spinal canal. An arterial plexus is thus formed around the head of each rib, from which a vessel is sent to each of the intercostal muscles and ribs and are continued into the muscles upon the outside of the body and its integuments. The anastomosis of the intercostal arteries round the ribs is similar to the plexus, which is produced by the great sympathetic nerve in the same location. The *lumbar arteries* are given off farther posteriorly and along the lumbar region. The lumbar, like the dorsal branches are given off in pairs. These arterial branches pass along the thigh and the upper sacral vertebral region, and supply those parts. Ventrally they are covered by the kidneys and pass into the abdominal muscles. The posterior aorta gives off the *spermatic arteries* and the *ovarian artery*. In the male the former supply the testes. The ovarian artery of the female gives off a twig to each calyx containing an ovum. Each calyx is voluminously supplied with blood. Next the *renal arteries* are given off to the kidneys.

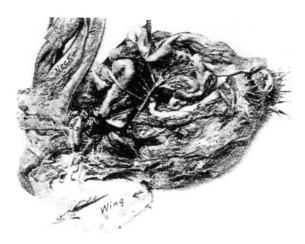

FIG. 63.—The vascular system injected. 1, The heart. 2, The posterior aorta. 3, The right brachio-cephalic artery. 4, The left brachio-cephalic artery. 5, The anterior mesenteric artery showing its many branches and anastomoses near and on the intestines. It is accompanied by branches of the mesenteric vein. 6, The rectal branch of the posterior mesenteric artery. 7, The duodenal loop and pancreas showing the pancreatic artery. 8, The anus. 9, The cloaca. 10, The liver. 11, The lungs. 12, The right subclavian artery. 13, The right carotid artery. 14, The right anterior vena cava. 15, The subclavian vein. 16, The right jugular vein. 17, The carotid trunk. 18, The posterior vena cava.

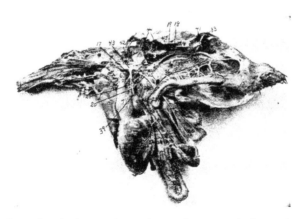

FIG. 64.—The splanchnic arteries, veins and nerves. 1, A portion of the left testes. 2, Adrenal gland. 3, Anterior lobe of kidney. 4, Heart. 5, Liver. 6, Second portion of the esophagus. 7, Proventriculus. 8, Gizzard. 9, Spleen. 10, Duodenal loop. 11, Pancreas. 12, Blind extremity of cæcum. 13, Floating portion of small intestine. 14, Rectum. 15, Cloaca. 16, Anus. 17, Celiac axis. 18, Ureter. 19, Vas deferens. 20, Recurrent sinister artery. 21, Anterior

recurrent dexter artery. 22, Arteria hepatica dextra. 23, Recurrent esophageal artery. 24, Recurrent intestinalis artery. 25, Ilio-colicus artery. 26, Anterior mesenteric artery. 27, Recurrent branches of the same. 28, Ilio-cœliacus artery. 29, Recurrent superior hemorrhoidal artery. 30, Posterior mesenteric artery. 31, Crural arteries. 32, Sacralis media artery. 33, Recurrent renalis. 34, Middle renal lobe. 35, Posterior renal lobe. 36, Vena mesentericus communis. 37, Vena hemorrhoidalis. 38, Vena pancreatico-duodenalis. 39, Vena proventriculo-leinealis. 40, Vena portalis dextra. 41, Ischiadic nerve. 42, Thoracic sympathetic trunk. 43, Anterior splanchnic plexus. 44, Spinal nerves. 45, Posterior splanchnic nerve plexus. 46, Intestinal nerve trunk. 47, Vena renalis magna.

The **celiac axis** (Fig. 64, No. 17) originates near the seventh sternal vertebra and to the right of the esophagus, breaks through the diaphragm, and, near this point, gives off a recurrent esophageal (Fig. 64, No. 23). The celiac axis gives off three main branches: The anterior or recurrent dexter, the posterior or recurrent sinister, the recurrent intestinalis, all of which lie to the right of the spleen, under the left lobe of the liver, and along the left side of the stomach. The *recurrent sinister* (Fig. 64, No. 20) gives arterial branches to the proventriculus, the gizzard, the pyloris, and the left lobe of the liver. The anterior or recurrent dexter (Fig. 64, No. 21) gives off a renal artery and from four to six splenic arteries.

The *recurrent intestinalis artery* (Fig. 64, No. 24) arises on the posterior of the stomach within the duodenal loop, and supplies the duodenum and the pancreas (Fig. 63, No. 7). It gives off a branch, called the recurrent ilio-colicus (Fig. 64, No. 25), which supplies the large intestines, including the cæca.

The *arteria hepatica dextra* (Fig. 64, No. 22) is a branch from the celiac axial trunk. It gives several branches to the right lobe of the liver and to the gall-bladder. The gastric branches pass to the muscles of the stomach. The anterior and posterior recurrent branches often anastomose.

The **anterior mesenteric artery** (Fig. 63, No. 5; Fig. 64, No. 26) originates from the posterior aorta near the generative glands. It is directed downward, and divides into many branches which pass in the mesentery toward the intestines. Recurrent branches (Fig. 64, No. 27) are given off, which anastomose, forming mesenteric arches. From these arches are given off branches which supply the intestinal walls.

The anterior mesenteric artery gives off the *recurrent ilio-celiacus* (Fig. 64, No. 28) which is distributed to the cæca; other of its branches are distributed to the small intestine.

The anterior mesenteric artery extends along the small intestine and ends in the *recurrent superior hemorrhoidal arteries* (Fig. 64, No. 29) which anastomoses on the surface of the rectum with a branch of the posterior mesenteric artery.

Branches from the anterior mesenteric artery also anastomose with branches of the celiac axis.

The **posterior mesenteric artery** (Fig. 64, No. 30) is given off from the posterior aorta posterior to the origin of the anterior mesenteric and near the arteries of the thigh and is distributed to the lesser mesentery supplying the large intestine. Branches of the superior hemorrhoidal arteries anastomose with branches of the posterior mesenteric artery. Branches from this source are called the *median hemorrhoidal* (Fig. 63, No. 6).

The posterior mesenteric artery (Fig. 64, No. 30) also gives off a long branch to the cæca. Branches of the posterior mesenteric artery anastomose with branches of the anterior mesenteric.

The **crural arteries** (Fig. 64, No. 31) are given off in a pair from the posterior aorta. They pass through the mass of the lobes of the kidneys, at about the median region, and extend out of the pelvic cavity in front of the spine and ilio-pubic ligament.

FIG. 65.—Blood-vessels and nerves of the hind extremity. Inside view of leg. 1, Ischiadic nerve. 2, Ischiadic artery. 3, Posterior tibial vein. 4, Vena metatarsa dorsalis interna. 5, Vena metatarsalis plantaris profunda. 6, Vena cruralis. 7, Vena metatarsalis interna, vena magna. 8, Vena tibialis postica. 9, Vena metatarsa magna. 10, Nerves of the toes. 11, Vena metatarsalis dorsalis profunda.

The crural, or the external iliac artery, is divided into three branches as follows:

First, the *internal pelvic*, or *umbilical artery*, is given off just as the cruralis leaves the pelvic cavity. It passes, as a long vessel, on the inner surface of the lumbo-sacral bones and supplies the obturator internus muscle and extends into the umbilical region. It gives off branches to the abdominal muscles, and, in the female, a branch is distributed to the ligament of the oviduct.

Second, the *arteria circumflex femoris*, which passes between the sartorius and vastus internus and extends dorsalward, supplies the sartorius muscle, the vasti muscles, and the ilio-trochanteric region.

Third, the *femoral artery*, which passes beside the vena cruralis and extends down the posterior surface of the thigh to the knee-joint, supplies the upper thigh muscles with the exception of the adductor muscles.

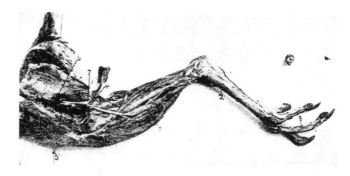

FIG. 66.—Blood-vessels and nerves of the posterior extremity. Outside view. 1, Anterior tibial artery. 2, Metatarsal artery. 3, Digital arteries. 4, Vena cutaneous crurus. 5, Ischiadic artery. 6, Vena cruralis. 7, Lateral cutaneous branch of the ischiadic nerve.

FIG. 67.—Blood-vessels and nerves of the fore limb. Outside view. 1, Median nerve. 3, Ulnar nerve. 4, Radial artery. 5, Recurrent ulnaris artery. 6, Twigs of ulnar artery to wing feathers.

The **ischiadic artery** (Fig. 69, No. 9) forms the main artery of the posterior extremity. This artery, the largest vessel of that region, is given off from the posterior aorta and passes ventrally over and between two of the main lobes of the kidney. The continuation of the posterior aorta is called the *sacralis media* (Fig. 64, No. 32). The ischiadic artery gives off a *recurrent renalis* on the posterior lobe of the kidney (Fig. 64, No. 33). On the left side it gives off a branch to the oviduct and to the ligament of the oviduct. The main trunk leaves the cavity with the ischiadic nerve (Fig. 65, No. 1) through a foramen formed by the os ilium and the os ischium. It sends branches into the adductor muscles of the upper and the lower thigh. It gives off anastomotic branches which unite with those from the femoral artery. At the flexure of the knee-joint it gives off two branches to that region (Fig. 69, No. 10). It gives off another branch to the gastrocnemius muscle and one to the flexor perforans digitorum. It terminates in the anterior and posterior tibial arteries.

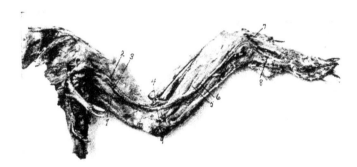

FIG. 68.—Blood-vessels and nerves of the fore limb. Inside view. 1, Vena humeri profunda. 2, Brachialis longus inferior. 3, Brachialis profunda. 4, Radial artery. 6, Ulnar artery. 7, Ulnar arterial branch to thumb. 8, Digital branch of ulnar artery. 9, Ulnar artery.

The first, the *tibialis postica* (Fig. 69, No. 12) passes between the gastrocnemius and the deep flexors giving off branches in its course to the skin and to other parts, and disappears shortly below the intertarsal joint.

The second, the *tibialis antica* (Fig. 66, No. 1; Fig. 69, No. 11), is often the larger of the two arteries. The anterior tibial artery gives off branches to the knee-joint and to adjacent structures. It gives a large branch to the head of the gastrocnemius muscle. It passes to the posterior surface of the tibial head and gives off the peroneal artery, which passes through the membrana interossea located between the tibia and fibula giving branches to the anterior side of the membrane of the patellar region of the knee-joint. It gives branches to the anterior side of the lower thigh and finally terminates subcutaneously in the anterior tibial plexus.

The main portion of the anterior tibial artery passes downward along the posterior surface of the membrane located between the tibia and fibula, and sends branches for the flexor digitorum communis et profundus. It then breaks through the membrane, reaches the anterior side and there communicates with the anterior tibial plexus. It gives off branches to the muscles on the anterior side of the lower thigh and the skin, and passes between the outer and the middle malleolus of the tarsal bones. It passes to the plantar surface and divides between the toes. In addition to the plantar branches there are given off dorsal branches between the third and the fourth toes. These branches form the direct continuation of the dorsal vessels of the anterior tibial plexus.

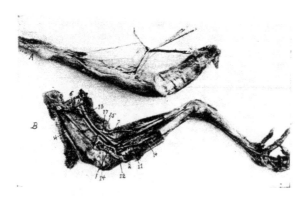

FIG. 69.—Blood-vessels and nerves of the femoro-tibial region. A section of the tibia removed.

A. Veins, arteries and nerves dissected as in *B*.

B. *V*, Vein. *A*, Artery. *N*, Nerve. 1, Vena poplitealis. 2, The three veins forming the vena poplitealis. 3, Vena femoris interna profunda. 4, Vena femoris anterior. 5, Vena cruralis. 6, Vena cutanea abdominalis femoralis. 7, Vena cutaneous cruralis. 8, Deep vein of the knee-joint. 9, Ischiadic artery. 10, Branches of ischiadic artery to flexure of knee. 11, Anterior tibial artery. 12, Posterior tibial artery. 13, Ischiadic nerve. 14, Trunk (a branch of the ischiadic) which gives off the superficialis peroneus and peroneus profundus. 15, Median branch of ischiadic nerve. 16, Superficialis peroneus. 17, Lateral branch of the ischiadic nerve.

The **arteria pudenda communis** (Fig. 70, No. 20) passes to the depressor coccygeus muscle, gives off a branch to the caudal part of the kidney, crosses the ureter, provides the ischio-pubic and pubio-coccygeus, or depressor coccygis lateralis, muscles, and on the lateral rim of the latter muscle it gives off the *arteria hemorrhoidalis intima*. This artery passes to the bursa of Fabricius and to the end of the cloaca. It enters into the ischio-coccygeus muscle, and divides into the *arteria pudenda externa*, and in ducks, the *arteria profunda penis*. These branches supply the vas deferens, the ureter, the cloaca, the penile structure, and the muscles of these parts.

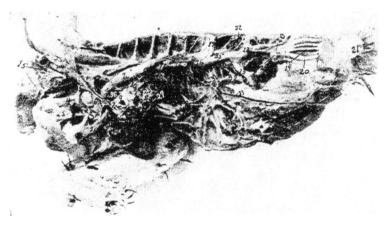

FIG. 70.—Blood-vessels of a Cornish cock. 1, The left pulmonary artery. 2, Right pulmonary artery. 3, Left brachio-cephalic artery. 4, Left subclavian artery. 5, Left carotid artery. 6, Right brachio-cephalic artery. 7, Posterior aorta. 8, Posterior vena cava. 9, Left pulmonary vein. 10, Right pulmonary vein. 11, Celiac axis. 12, Anterior mesenteric artery. 13, Ischiadic artery. 14, Crop. 15, Superior artery of the crop and vein of same name. 16, Testicular or ovarian artery. 17, External iliac artery. 18, Intercostal nerve. 19, Median sacral artery. 20, Arteria pudenda communis. 21, Anus. 22, Vena hypogastrica. 23, Right vena hepatica magna. 24, Left vena hepatica magna. 25, Vena iliaca interna. 26, Vena iliaca communis. 27, Vena iliaca externa. 28, Vena coccygo-mesenterica. 29, Vena umbilicalis. 30, Vena suprarenalis. 31, Origin of the pars renalis. 32, Lumbar veins. 33, Vena epigastrica. 34, Lumbales (arteries). 35, Anterior division of the lumbo-sacral plexus. 36, Posterior division of the lumbo-sacral plexus. 37, Left anterior vena cava. 38, Right anterior vena cava.

The Sacralis Media.—The *median coccygeal artery* (Fig. 70, No. 19) forms the single extension of the sacralis media. It gives off lateral branches between the caudal vertebræ which supply the dorsal muscles of that region and the skin. The third pair are the largest. These are the *arteriæ coccygeæ laterales*, and are located on the dorsal side of the tail, they supply the tail glands and rudder feathers, the main tail feathers, or rectrices. A small arterial twig is given off to the papilla of each rectrix.

THE VENOUS TRUNKS

The venous blood enters the lungs from the right ventricle through the two pulmonary arteries (Fig. 70, No. 1 and 2).

The two pulmonary veins (Fig. 70, No. 9 and 10) collect the arterial blood from the lungs, and empty it into the left auricle.

There are three venæ cavæ which collect the systemic blood and empty it into the right auricle. These venæ cavæ are two anterior (Fig. 70, No. 37 and 38) and one posterior (Fig. 70, No. 8).

Each anterior vena cava is formed by the union of a vena jugularis, a vena vertebralis, and a vena subclavicularis (Fig. 63, No. 15).

FIG. 71.—Veins of the liver of a fowl. 1, Vena mesenterica communis. 2, Vena portalis propria. 3, Anterior vena mesentericus. 4, Vena portalis dexter. 5, Vena portalis sinister. 6, Posterior vena cava. 7, Celiac axis. 8, Base of the heart. 9, The liver. 10, Hepatic veins.

The jugular vein is formed by the union of the vena cephalica anterior, and the vena cephalica posterior.

The jugular vein (Fig. 62, No. B, 3) passes along the side of the neck and lies near the trachea, the esophagus, and the pneumogastric nerve. Near the base of the skull the two jugular veins, the right and the left, are connected by a transverse vein. By this anastomosis part of the blood from the left jugular vein is sent into the right. Therefore the right jugular vein is larger than the left. The jugular veins collect the blood from the tongue region, the thyroid, the esophagus, the trachea, the crop, and other structures along its course through the cervical region.

The **vertebral veins** are divided into the anterior and the posterior, or the inferior and the superior. The *anterior vertebral vein* is located in the cervical region and collects the blood from the brain and the inner part of the head. The vertebral vein passes along the dorsal side of the spinal cord.

BRANCHES OF THE VENÆ CAVÆ ANTERIORES

(Venæ linguales

(Vena occipitalis lateralis (Vena occipito-collores

(Vena ascendenes lateralis

(Venæ colli cutineæ

(Venæ esophagealeæ

(Venæ tracheales

(Vena   (Vena subscapularis

(jugularis (Venæ glandularum thyroidearum

(((Vena occipitalis (Sinus foraminis

(((interna   (occipitalis

((Vena   (

((vertebralis (Vena vertebralis posterior (Vena intercostales

Vena cava (((Vena vertebralis anterior

anterior ((Vena vertebralis

(Vena vertebralis lateralis dorsalis (Venæ intercostales

(

(Venæ coronariæ

(Vena subclavicularis

(Vena thoracica interna (Vena intercostales
(Vena pharyngea superior
(Vena muscularis depressoris mandibularis
Vena facialis (Vena muscularis colli anterior superioris
communis   (Vena lingualis
  (Vena sublingualis et sphenoidea

The *posterior vertebral vein* is located in the dorsal region, passes backward and receives the blood from the dorsal neck region and from the intercostal veins and from the vertebral segments and adjacent regions.

The anterior and the posterior vertebral vein form one trunk, the vertebral vein, and this trunk empties into the vena jugularis just before the subclavian and the jugular unite.

The **subclavian vein** collects the blood from the anterior extremities. It unites with the vena jugularis of the same side.

The left vena cava receives the coronary veins from the heart.

THE VEINS OF THE HEAD
BRANCHES OF THE VENA FACIALIS INTERNA

The **vena facialis interna** (Fig. 62, No. 12) lies dorsal on the pterygoid bone and receives the vena maxillaris, the vena ophthalmica, the vena mandibularis interna, the venæ pharyngeæ superiores, and the vena retis mirabilis temporale.

FIG. 72.—Blood-vessels and nerves of the abdominal cavity and hind extremity and head. 1, Vena renalis. 2, Vena intervertebrales sacrales. 3, Posterior aorta. 4, External iliac. 6, Ischiadic nerve. 7, Ischiadic artery. 8,

Arterial rete (plexus temporalis). 9, Ophthalmic artery. 10, Palatine artery. 11, Lingual artery. 12, Hyoid artery. 13, 14, External carotid artery. 15, Internal maxillary artery. 16, Superior laryngeal artery. 17, Superior esophageal artery. 18, Occipital artery. 19, Internal carotid artery.

The **vena maxillaris** comes out of the upper beak with the recurrent trigeminus nerve and then passes backward between the jugular and palatine bones. It lies medio-ventrally to the eyeball and communicates at this point with the ophthalmic vein. It is covered ventrally with the pterygoid bone. The maxillary vein collects the blood from the gland of Harder, the upper beak, and receives blood from the *vena supra-palatina* and from the lower jaw bone and the external mandibular vein. It also receives blood from the external sublingual vein, and the sublingual gland. It follows the inner rim of the jaw bone. The maxillary vein receives an anastomosing branch from the cutaneous facial vein, at a point near the commissure of the mouth.

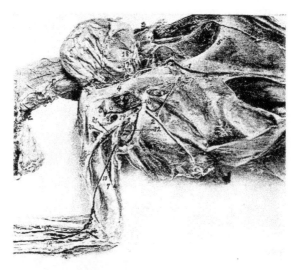

FIG. 73.—Photograph of blood-vessels and nerves of the thorax. 1, Vena sternalis. 2, Vena thoracico-externa. 3, External thoracic artery. 4, Coracoid bone. 5, Sternum. 6, Shoulder-joint. 7, Humerus. 8, Clavicularis artery. 9, Sterno-clavicularis artery. 10, Thoracico-humeralis artery. 11, Subclavian artery. 12, Axillary artery. 13, Brachialis profundus artery. 14, Ulnar artery. 15, Radial artery. 16, Internal mammary artery. 17, Superior vein of the crop. 18, Superior artery of the crop. 19, Brachial nerve plexus. 20, Anterior division of brachial nerve plexus. 21, Posterior division of brachial nerve plexus.

The **vena ophthalmica** lies close to the point where the olfactory nerve passes out of the cranial cavity. The largest branch it receives is the recurrent ophthalmo-temporalis, which accompanies the lateral side of the optic nerve. It collects the blood from the base of the brain, from Harder's gland, from the eye muscles, from the inner eye, the nose, the lacrimal gland, the skin of the frontal region, and the comb.

BRANCHES OF THE VENA JUGULARIS

```
   ( Vena maxillaris

      ( Vena ophthalmica
( Vena facialis   ( Vena mandibularis interna
( interna        ( Vena retis mirabilis temporalis
( Vena cephalica (          ( Venæ pharyngeæ superiores
( anterior or    (


( facialis      (         ( Vena facialis cutinea
( communis      (         ( Vena temporalis


(         ( Vena facialis   ( Vena auricularis

Vena jugularis (         ( externa      ( Vena palpebralis

or cephalica  (                    ( Vena transversus
communis      (                    ( Vena retis mirabilis temporalis
( Vena subclavia
( Venæ vertebrales
(         ( Sinus cranii et cerebri
(         ( Vena carotis
```

((

((Vena auris   (Sinus foveæ cerebelli
(((Vena occipitalis externa

(Vena cephalica   (Vena jugularis prima

(posterior   (Vena occipitalis

(Vena lingualis

(Vena mandibularis externa

Vena maxillaris (Vena sublingualis externa

(Vena supra-palatina

Vena ophthalmica (Vena ophthalmo-temporalis, recurrent

Vena facialis communis (Vena cephalica externa

Vena occipitalis externa (Vena occipitalis interna

Vena retis mirabilis temporalis (Vena spheno-temporalis

Sinus foveæ cerebelli (Sinus occipitalis externa

The trunk of the internal facial vein receives the blood of the **internal mandibular vein**. The internal mandibular vein emerges from the alveolar canal of the inferior jaw bone and receives branches from the muscles of mastication.

The **vena retis mirabilis temporalis** passes out of the eye cavity and lies between the internal facial vein and the external facial vein, and passes around the os quadratum and the os pterygoideum. It receives veins from the outer ear region, lacrimal gland, eyelids, and the spheno-temporalis.

The **venæ pharyngeæ superiores** are the veins on the dorsal surface of the pharynx. These veins form the small plexuses which unite with the two trunks of the facial veins. This forms an anastomosis between the left and the right facial veins.

BRANCHES OF THE EXTERNAL FACIAL VEIN

The vena facialis externa (Fig. 62, No. B, 9) lies behind the quadrate bone and is partly covered ventrally by the inner extension of the lower jaw bone. It collects the blood principally from the upper region of the cranium and the face, including the comb.

The **vena facialis cutinea** receives branches from the muscles in the region of the jaw, the lower jaw bone, the muscles of the eyelids, the frontal region, and the anastomotic branches from the vena maxillaris.

The **vena temporalis** is made up of veins from the skin of the temporal region, the masseter and the tongue muscles, and from the sides of the upper throat region.

The **vena palpebralis** is made up from veins from the three eyelids and the temporal region near the eye. Its trunk passes laterally over the temporo-mandibularis ligament. The vena palpebralis collects blood from the lacrimal gland, the eyelids, and the outer ear region.

The following veins empty into the trunk of the **vena facialis communis**: venæ pharyngis superiores, vena muscularis depressoris mandibulæ, vena muscularis colli anterior superioris, vena lingualis et hyoidea. The right terminal branches of the vena lingualis empty into the vena cephalica posterior.

The venæ pharyngis superiores and vena lingualis et hyoidea collect blood from the muscles of the tongue and posterior tongue region.

The *vena lingualis et hyoidea* (Fig. 62, No. B, 7) collects the blood from the muscles of the tongue region, the lower tongue glands, and the upper throat region.

Between the two trunks of the vena facialis communis and anterior to the entrance of the vena cephalica posterior there is an anastomosis, the vena transversus (Fig. 62, No. B, 5). Through this anastomosis which lies crosswise the two jugular veins are influenced in different ways. At this point both jugular veins are of the same size. The anastomotic branch lies crosswise. Through this anastomosis the right receives some blood from the left head region indicating that the blood flows from left to right. This, as stated before, makes the left jugular vein the smaller.

BRANCHES OF THE VENA CEPHALICA POSTERIOR

The left vena cephalica posterior (Fig. 62, No. B, 8) communicates with the transverse vein (Fig. 62, No. B, 5). The left receives the smaller veins of the skin region, the esophagus, and the trachea. These vessels communicate with their fellow of the opposite side and with the vertebral vein of the same side, which causes a gradual reduction of the size of the jugular vein.

BRANCHES OF THE VENA OCCIPITALIS EXTERNUS

```
(              ( Venæ choroideæ

( Sinus        ( Venæ cutaneæ et frontales

( longitudinalis  ( Venæ nasales

Vena           (           ( Venæ ophthalmiæ
occipitalis (

externus  (            (         ( Sinus occipitalis
( Sinus        ( Sinus   ( Sinus temporo-sphenoideus

( semicircularis ( transversus ( Venæ cerebellares
(              (         ( Sinus petrosus sphenoideus
```

Sinus petrosus sphenoideus (Retis mirabile temporalis

Sinus temporo-sphenoideus (Sinus annularis venosus basilaris

Sinus foraminus occipitalis (Venæ medullares

(Venæ cerebrales

Sinus (Vena cerebralis basilaris (Sinus venosus annularis anterior

venosus (

annularis (

basilaris (Vena ophthalmica

(Vena medullaris mediana longitudinalis

The vena cephalica posterior (Fig. 62, No. B, 8) is formed from the veins of the sinuses of the cranium, the brain, the vena auris, the vena jugularis prima, the vena carotis, and the vena occipitalis. It collects most of the blood from the cranium and the posterior part of the head and from the tongue.

THE VENOUS SINUSES OF THE HEAD

The **sinus longitudinalis** (Fig. 62, *A*, 10) extends from the dorsal median line of the internal occipital protuberance to the olfactory nerve. It collects the blood from the choroidea of the brain and from the skin of the frontal region. It receives laterally the veins of the nose, the vena ophthalmica, and the sinus transversus.

The **sinus occipitalis** (Fig. 62, No. *A*, 4) forms the continuation of the transverse sinus. It is located in the posterior cerebral cavity, and is wing-shaped, extending both to the right and to the left.

The **sinus foraminis occipitalis** a continuation of the sinus occipitalis, lies transversely on the ventral side of the cranial cavity and receives the veins from the medulla oblongata. One branch passes through the os occipitalis basilare, and then extends sidewise to the foramen magnum and anastomoses with the vena occipitalis.

The **sinus transversus** (Fig. 62, No. *A*, 6) extends in a pair from the internal occipital protuberance between the cerebrum, cerebellum, and the corpus quadrigeminum. It receives vessels from the upper surface of the cerebellum. The sinus has three branches, all of which empty into the sinus semicircularis.

The **sinus semicircularis** extends along the dorsal petrosal rim, anterior to the inner ear, and from this point backward. Near the foramen magnum it passes through the os occipitale and empties into the external occipital vein.

The **vena occipitalis externus** extends near the foramen magnum, surrounds the basis occipitalis, and receives the semicircular sinus and through the semicircular sinus the blood from the sinus transversus.

The **sinus petrosus sphenoideus**, passing in company with the occipital sinus, extends from the transverse sinus to the cavity near which the trigeminus lies. The sinus dividing into two branches, one of these extends to the hypophysis and unites with the basal veins of the brain; the other passes with the rete out of the cranial cavity along with a branch of the trigeminus, and communicates with the rete mirabile temporale.

The **sinus temporo-sphenoideus** (Fig. 62, No. *A*, 5) extends forward and outward from the transverse sinus, and, between the corpus quadrigeminum and the cerebrum, unites with the sinus venosus annularis basilaris.

The **sinus venosus annularis basilaris** lies in the region of the optic nerve and surrounds the hypophysis, forming a complete circle. It receives small vessels from the cerebrum, the corpora quadrigemina, and anteriorly the longitudinal basilar vein or vena basilaris media, of the cerebrum, which comes out of the anterior annular venous sinus. The anterior annular venous sinus surrounds the base of the olfactory nerves. It communicates with the sinus petrosus sphenoideus, which emerges from behind the optic lobe, the sinus temporo-sphenoideus, which emerges from in front of the optic lobe, and posteriorly the median longitudinal vein of the medulla oblongata or vena basilaris. Laterally and anterior to the optic lobe it receives the vena basilaris lateralis, and also a branch of the vena ophthalmica.

The blood received by the **sinus annularis basilaris** comes from the brain cavity through three pairs of veins, as follows: first, the two ophthalmic veins which pass through the posterior orbital wall between the olfactory and the optic nerve; second, two other branches of the ophthalmica which pass with the optic nerve; third, two veins which pass through a foramen by the side of the sella turcica to the hypophysis together with the carotis cerebralis, and then leave the cerebral cavity ventrally, at which point they are called the venæ carotes. These last accompany the cerebral artery backward, and pass through a foramen in the base of the cranium. Each one then passes through the cranial wall close to the external auditory canal, and empties into another vein in the posterior region of the head.

The **sinus fovæ hemispherii cerebelli** lies in the hollow between the os petrosum and the semicircular canals. It receives blood from the external occipital sinus and from the cerebellum, and empties into the vena auris interna.

VEINS OF THE BRAIN CAVITY

The **vena auris interna** passes through a bony canal along the outer rim of the posterior semicircular canal and then along the external semicircular canal. It receives vessels from the labyrinth, and extends along the posterior rim of the tympanum to the outer ear canal.

The **vena occipitalis interna**, extending from the sinus foraminis occipitalis, passes through the side of the atlas and divides near the condyle into two branches. One of these branches the internal occipital, forms the root of the vena vertebralis, and the other empties into the vena occipitalis externa.

Near the outer and upper part of the condyle it forms two large veins and collects the blood from the vena occipitalis media. These veins collect the blood from the rectus capitis anticus, and communicate with the vena cephalica anterior.

The **vena occipitalis externa**, sometimes called the vena occipitalis collateralis, extends from the vena auris, and receives vessels from the neck region. It receives on the side of the occipitoatloid joint the vena occipitalis interna. The vena occipitalis externa forms a junction with the vena vertebralis emptying into the transverse vein.

THE VEINS OF THE NECK

BRANCHES OF THE VENA JUGULARIS

The trunks of the vena jugularis or vena cephalica communis pass subcutaneously on both sides near the trachea and the esophagus. Near the height of the thyroid gland they almost touch the carotid artery. They then cross, dorsalward, the trunk of the subclavian artery, and in their course receive the vertebral vein and unite with the subclavian vein toward the side of the vena brachiocephalica. The union on the right side lies to the right of the anterior aorta, and the union on the left side lies to the left of the pulmonary artery.

The jugular vein (Fig. 18, No. 4) receives in its course the following veins:

The **venæ linguales** (Fig. 62, No. B, 7), which at times empty into the vena cephalica posterior.

The **vena occipito-collares**, which arises on the ventral side from the muscles of the neck and the vertebræ; empties into the vena occipitalis lateralis and also communicates with the vena jugularis by anastomoses.

The **venæ ascendentes laterales**, which collect blood from the lateral sides of the neck.

The **venæ colli cutineæ**, which enter the jugular vein laterally. The upper branches are directed crosswise, and are shorter than the lower ones. The lower branches are located near the thorax, are directed upward, and anastomose by fine terminal twigs. Plexuses are frequently formed in the inner side of the skin of the neck.

BRANCHES OF THE VENA SUBCLAVIA

(Vena profunda ulnaris

(Vena brachialis (Vena profunda radialis
(Truncus venæ ((Vena profunda humeri

(axillaris   (

((Vena basilica or cutinea ulnaris

(

((Vena cutinea abdomino-pectoralis

Truncus venæ (Vena pectoris (Vena infrascapularis anterior

subclaviæ (externa   (Vena thoracica externa

(Vena coracoidea

(Vena sternalis

The **venæ esophageæ** are numerous smaller veins formed in closely woven plexuses in the region of the esophagus along the neck. Some pass upward and some downward.

The **venæ tracheales** are located on the supero-lateral side of the trachea and along the edge of the lateral tracheal muscle. They have numerous anastomoses transversely with the longitudinal vein of the cervical region, anteriorly with the vena lingualis, and on the left side with the left jugular vein.

The **vena subscapularis** passes along the median line of the scapula to the side of the jugular vein, and anastomoses with the vertebral vein.

The **venæ glandularum thyroidearum** are several small, short veins coming from the thyroid gland.

The **vena vertebralis**, originating near the atlas, is a continuation of the internal occipital vein. The internal occipital vein receives the bulk of the blood from the brain and communicates with the lateral veins of the cervical region and collects blood from the cervical vertebra. The vertebral vein accompanies the vertebral artery and the deep sympathetic nerve trunk in the vertebral canal. The vein either leaves this canal with the vertebral artery at a point near the last two vertebral nerve trunks of the brachial plexus, or it leaves the canal one vertebra in front of the vertebral artery.

THE VEINS OF THE DORSAL REGION

The **vena vertebralis lateralis dorsalis** collects blood from the cervical vertebræ, from the dorsal vertebræ, and from the intercostal region.

The blood from the neural canal, including the spinal cord, is collected in one long vein extending from the head to the tail on the upper side of the cord. This vein has sinus-like expansions. Between each two vertebræ it anastomoses with the vertebral vein on each side. In the lumbo-sacral region these anastomotic branches empty into the hypogastric veins.

The **vena intercostalis** is formed in the costal region and anastomoses with the vein extending longitudinally between the capitulum and the tuberculum of the rib.

The **vena vertebralis posterior** at the height of the first and the second dorsal vertebræ, empties into the main trunk of the vertebral vein. Its lower roots come out near the side of the kidney, and unite, forming a large ascending trunk uniting with the vena vertebralis anterior forming one large trunk, which extends into the chest cavity.

The posterior vertebral vein also receives a large vein from the skin of the outer tarsal region and a few vessels from the muscles of the outer abdominal and the outer costal region, from the skin, from the pectoralis muscles, and from the shoulder region.

VEINS OF THE THORAX

BRANCHES OF THE VENA SUBCLAVIA

The **vena thoracica externa** (Fig. 73, No. 2) is made up of veins mainly from the pectoral group of muscles. The posterior branch anastomoses with the vena cutinea. Its anterior branch is made up mainly of branches from the furcular region. It empties into the vena pectoralis externa and the vena pectoralis externa empties into the vena subclavia.

The **vena coracoidea** comes out of the region of the shoulder-joint, passes downward along the inner surface of the coracoid, and receives small branches from the walls of the arteria brachia cephalica and from the pericardium.

The **vena sternalis** (Fig. 73, No. 1) is made up of two branches. The outer branch comes from the muscles of the subclavian region passes over the cristi sterni, medially, to the sterno-coracoid joint and into the breast cavity, where it receives the inner branch, which drains the inner surface of the breast-bone.

The **vena thoracica interna** empties into the vena cava anterior. It extends from the abdominal muscles where it communicates with the epigastric vein. It then passes on the inner side of the thoracic cavity close to the breast-bone and receives many intercostal veins.

The **vena cava sinistra** receives first, the *vena proventricularis communis*, which collects the blood from the walls of the proventriculus, and second, the vena coronaria cardis magna. This latter vein originates close to the apex of the heart and collects blood principally from the walls of the left ventricle. It connects in the left sulcus transversus with the vena cardis superior and ends at the base of the left upper vena cava. Its exit is not guarded by a valve. The veins of the right ventricle are partly on the surface. They collect blood along the sulcus transversus dexter and enter directly into the right ventricle. The veins of the front part of the heart are small. They collect behind the sulcus coronalis and end either directly into the right auricle or into the vena cardis magna.

VEINS OF THE FORE LIMB

The **vena radialis profunda** accompanies the radial artery on the dorsal anterior rim of the index-finger and passes, on the dorsal side, over the carpal region. It passes through the interosseous ligament between the ulna and the radius, and reaches the ventral surface of the arm. At this point it passes upward and anastomoses with the vena ulnaris. It collects the blood from the skin of the anterior wing region and the flexor muscles of the anterior arm. It empties into the vena brachialis.

The **vena humeri profunda** (Fig. 68, No. 1) emerges at the height of the elbow, and collects blood from the skin of the dorsal surface of the wing. It also receives veins from the muscles of the posterior side of the upper arm. It passes subcutaneously and dorsally over the dorsal portion of the humerus in company with the external radial muscle and passes with it around the external part of the humerus between the long and short heads of the triceps. It takes a diagonal course to the shoulder cavity and at that point empties into the brachial trunk.

The **profundus ulnaris** originates at the volar surface of the hand, proceeds in company with the ulnar artery, and sends on the base of the hand small anastomotic veins to the vena cutanea ulnaris or basilica. It passes along the anterior arm and between the flexor carpi ulnaris and the pronator profundus muscle to the elbow-joint. On the median surface of the biceps it passes upward and anastomoses with the radial vein. The ulnar vein, in the region of the elbow-joint, receives a large lateral branch which extends around the end tendon of the biceps and anastomoses above with the vena basilica.

The **vena basilica** or cutanea ulnaris is a long vein which originates from the subcutaneous dorsal surface of the index-finger. Near the base of the hand it receives an anastomosing branch from the vena radialis and the vena ulnaris, and then passes upward along the posterior rim of the ulna. It receives numerous branches from the roots of the flight feathers. It crosses below the elbow-joint and reaches the volar surface of the arm. It receives a large branch from the ulnar vein, and then, passing in a diagonal and median direction to the triceps, extends to the shoulder cavity where it empties into the axillary vein.

The outer breast veins unite forming a trunk which crosses ventrally to the subclavian artery, and empties into the subclavian vein.

The vena cutanea abdomino-pectoralis collects the blood from a large skin area of the abdomen, the upper thigh, the breast, and the intercostal region. In the skin of the abdomen it forms a network.

BRANCHES OF THE ILIACS

(Vena metatarsalis   (Vena metatarsalis

(Vena   (interna or magna   (plantaris profunda

(tibialis (Vena tibialis antica

(postica   (Vena metatarsalis   (Venæ metatarsales

((externa

(

(Vena   (Vena metatarsalis dorsalis profunda

(Vena   (tibialis (Vena metatarsalis dorsalis interna

(poplitealis (antica   (Vena peronealis

((Vena cutanea cruralis

((Venæ surales
Vena   (

iliaca   ((Vena cutanea abdominalis femoralis
externa (Vena   (Vena femoralis interna profunda

(cruralis (Vena femoralis anterior

((Vena epigastrica

The main trunk passes in the median abdominal line forward and then upward, along the outer edge of the pectoralis major, over the first sternal rib; receives blood from the infrascapular vein, and empties into the pectoralis externa.

The **vena brachialis** is located in the middle of the humerus and the triceps muscle. It is formed by the vena ulnaris, and the vena radialis. It passes with the median nerve and the brachial artery over the inner surface of the humeral joint. Posterior to the humeral head it receives the vena profunda humeri. At the shoulder cavity it receives the vena cutanea ulnaris or vena basilica.

The trunk of the axillary vein is very short and is formed by the veins of the shoulder and the wing. The deeper wing veins accompany the large arterial and the nerve trunks.

BRANCHES OF THE VENA ILIACA INTERNA

(Vena intervertebralis lumbalis

(Pars truncalis (Vena renalis magna
((

(((Vena hypogastrica caudalis sinistra

((Vena renalis (Vena hypogastrica caudalis dextra

(((Vena portalis

((Vena coccygea (Vena cutanea et pudenda

Vena iliaca   ((Vena coccygea mesenterica

interna or   ((Vena cutanea pubica

vena hypogastrica (Pars caudalis (Vena cutanea caudalis

((Vena pudenda (Vena spermatica

((Vena caudalis muscularis

((Venæ sacrales
((Venæ intervertebrales sacralis (Venæ renales
((Venæ renales

(Pars renalis (Vena ischiadica
(Vena obturatoria
(Vena suprarenalis externa (Azygos sacralis

THE POSTERIOR VENA CAVA

The posterior vena cava (Fig. 63, No. 18; Fig. 70, No. 8) has its origin in the posterior half of the body of the bird, somewhat to the right of the posterior aorta, near the anterior lobe of the kidney, by the union of the right and the left vena iliaca communis. It receives the blood from all of the posterior half of the body including the posterior limbs, of the visceral organs, of the abdominal and the pelvic cavity. The posterior vena cava passes dorsally through the right lobe of the liver and through the diaphragm and ends in a short, broad trunk on the posterior dorsal side of the right auricle of the heart. Its opening into the heart is guarded by two half-moon shaped valves. The basal part of the trunk reaches from the right auricle of the heart to the upper anterior rim of the liver where it receives three large trunks, first, the right and second, the left vena portalis hepatica magna, and third, smaller vessels from the liver substance.

THE VEINS OF THE POSTERIOR EXTREMITY

In the skin region on the inner sides of the toes near their bases the small veins collect into five metatarsal veins. The largest vein collects the blood from the first, the second, and the third toe, passes up the tarsus and is called the vena metatarsalis interna, or vena magna (Fig. 65, No. 7). It is located just beneath the skin on the inner surface of the metatarsal bone. It passes in a circle, around the condyle of the tibia and becomes the **vena tibialis postica** (Fig. 65, No. 3 and 8). The vena tibialis postica passes under the tendon Achillis and the tendon of the flexor digitorum brevis, lies subcutaneously upon the latter, and reaching the knee-joint crosses over the upper surface of the ischiadic nerve and becomes the **vena poplitealis**, at which point it receives the vena tibialis antica.

BRANCHES OF THE VENA CAVA POSTERIOR

(Vena ovariana

(Venæ testiculæ

(Vena proventricularis communis

(Vena suprarenalis revehentis

(Vena portalis magna sinistra

(Vena portalis magna dextra

(Venæ innominatæ

(Venæ hepaticæ

Vena cava (Vena cardis coronaria magna
posterior (Vena proventricularis inferior

((Vena iliaca interna (Vena hypogastrica

((Vena iliaca externa (Vena renalis
((Vena suprarenalis externa
(Vena iliaca (Vena ischiadica

(communis   (Vena renalis
((Vena intervertebralis lumbalis

((Vena renalis magna

(Vena mesenterica communis (Vena coccygo-mesenterica

Vena   ((Vena mesenterica anterior

portalis (Vena pancreatico-duodenalis
dextra ((Vena proventricularis
(Vena proventricularis lienalis (Vena splenica

Vena coccygo-mesenterica (Vena hemorrhoidalis

On the dorsal side of the metatarsus are two veins. The **vena metatarsalis dorsalis profunda** (Fig. 65, No. 11), which extends under the tendon of the extensors of the toes, along with the artery and the nerve. It collects the blood from the third and the fourth toe and in the middle of the metatarsus receives the **vena metatarsalis dorsalis interna** (Fig. 65, No. 4), which connects the vena metatarsalis dorsalis profunda and the vena magna. The two dorsal veins anastomose with the vena magna, at the intertarsal joint. They pass transversely under the ligament and form the main trunk of the vena tibialis antica which lies close to the anterior surface of the tibia. Near this point there is formed a plexus of veins which again form a trunk and communicates with the vena peronealis and enters between the tibia and fibula with the tibialis antica. It extends along the flexure of the knee and the posterior part of the lower thigh.

The **vena metatarsalis externa** passes subcutaneously on the outside of the fourth toe and the metatarsus, and above the intertarsal point joins the tibialis postica.

The **vena metatarsalis plantaris profunda** (Fig. 65, No. 5) lies on the ventral side of the foot and forms several anastomosing arches with the other veins of the toes. Below the intertarsal joint it enters the vena metatarsalis magna (Fig. 65, No. 9).

The **vena cutanea cruralis** (Fig. 69, No. 7) originates at the height of the tarsal region and passes subcutaneously on the outer posterior surface of the lower thigh region to the vena poplitealis.

The **venæ surales** or inferior muscular branches of the vena poplitealis consist of many veins. One branch comes from the region of the shank and from the gastrocnemius muscle; another as a main branch from the posterior surface of the lower thigh; and a third from the outer surface of the muscles and skin of the upper thigh. The three branches together with the anterior and posterior tibial (Fig. 69, No. 2) unite at the flexure of the knee forming the vena poplitealis (Fig. 69, No. 1).

In the region of the upper thigh, between the knee and the abdominal cavity, the following four veins form the vena cruralis: (Fig. 66, No. 6; Fig. 69, No. 5). First, the **vena cutanea abdominalis femoralis** (Fig. 69, No. 6) which comes out of the side of the abdominal wall, draining the skin of the inner surface of the upper portion of the thigh, the adductor muscles, and the region of the abdominal and the breast border. It crosses the ischiadic artery in a diagonal direction, and enters the vena cruralis in the middle of the crural region.

Second, the **vena femoralis interna profunda** (Fig. 69, No. 3) forms a communication between the end of the suralis near the knee. It lies on the median portion of the flexor cruris internus muscle.

Third, the **vena femoralis anterior** (Fig. 69, No. 4) is formed from branches from the sartorius and adjacent structures, and anteriorly empties into the crural vein near where the latter enters the abdominal cavity.

Fourth, the **vena epigastrica** (Fig. 70, No. 33) is formed by branches from the abdominal wall and branches from the walls of the abdominal air-sacs. It passes along the median surface of the os pubis and ends into the vena cruralis near the spine of the ilio-pubica or at a point where these join with the vena hypogastrica.

VEINS OF THE CAUDAL REGION AND OF THE PELVIC CAVITY

The vena iliaca interna or the vena hypogastrica (Fig. 70, No. 22) collects most of the blood from the tail. The vena iliaca interna collects most of the blood from the pelvic cavity, and the adjacent intestines. It unites with the vena iliaca externa (Fig. 70, No. 27) and receives the vena renalis magna and forms the trunk of the vena iliaca communis (Fig. 70, No. 26).

The **vena coccygea** originates between the coccygeal vertebræ, and collects the blood from the region of the tail, including the tail feathers, the tail muscles, the tail gland, and the skin of the region. These small collecting vessels form a trunk on each side of the coccyx. The right and the left pass laterally, each one taking up a vena cutanea et pudenda, and frequently anastomosing with the vessels on the other side. These often unite into one vessel. Both trunks are connected by a transverse, or anastomotic, vessel. At this anastomosis there empty into it the vena coccygea mesenterica and the vena portalis. There also communicate at this point the right and the left hypogastrica caudalis. In its course it is partly imbedded in the kidney and passes anteriorly to the vena iliaca communis. This circle is called the arcus hypogastricus. Thus the veins of the abdominal cavity have many anastomoses forming many arcs, making possible two outlets for the blood.

The azygos sacralis empties into the arcus at about its middle. From here it extends forward to the inside and under the kidney and empties into the vena suprarenalis externa.

The **vena cutanea pubica** originates on the lower portion of the abdomen, collects the blood from the muscles of the distal part of the ischium, and enters the pelvic cavity between the ischium and the ilium. It joins with the vena cutanea caudalis.

The **vena cutanea caudalis** originates from branches which drain the skin and other parts of the ventral coccygeal region. The vena cutanea pubica also communicates with the vena caudalis muscularis and with the vena pudenda, thus forming the caudal trunk of the vena hypogastrica.

The **vena pudenda** originates in the walls of the cloaca in the region of the generative organs.

A small **vena spermatica** accompanies the lower end of the vas deferens and the ureter, and empties medially into the vena pudenda.

The pars renalis of the vena hypogastrica extends from the middle of the arcus to the union of the vena hypogastrica and the vena cruralis.

The **vena hypogastrica** communicates with the pars renalis. The vessels that empty into the pars renalis are as follows:

The **venæ sacrales** collect blood from the dorsal wall of the abdominal cavity and enter the pelvis through the foramen sacralis. They pass between the pelvic wall and kidney, and at times pass through the kidney tissue. They empty into the pars renalis of the hypogastric arch.

The **venæ intervertebrales sacrales** (Fig. 72, No. 2) originate in the region of the roots of the plexus of sacral nerves and pass through the kidney substance or on the dorsal surface and empty into the pars renalis.

The **venæ renales** (Fig. 72, No. 1) are very numerous and originate in the kidney substance, forming two main and several minor branches, which pass posteriorly, and empty partly into the vena hypogastrica, partly into the vena renalis magna, and also into the vena intervertebralis; other branches empty into the trunk of the iliaca communis.

The **vena suprarenalis externa** (Fig. 70, No. 30) is located near the anterior rim of the kidney and is connected with the vena hypogastrica. It also receives on the medial side, short branches which come out of the upper kidney surface and sacral vertebræ.

The **vena ischiatica** originates by the union of several venous branches which come from the muscles of the pelvis and upper thigh region. The ischiatic vein enters the pelvic cavity along with the ischiatic nerve and artery, and communicates with the vena hypogastrica at about the level of the anterior lobe of the kidney. It is always smaller than the vena cruralis.

The **vena obturatoria** originates mainly from vessels from the obturator muscles. It enters the pelvic cavity through the obturator foramen. Another branch is sometimes found which comes out of the inner surface of the peritoneum which covers the obturator muscles and the walls of the abdominal air-sacs and empties between the vena ischiatica and the vena vertebralis into the vena hypogastrica.

VEINS OF THE TRUNCUS VENA ILIACA COMMUNIS

The **vena intervertebralis lumbalis**, which comes out of the lumbar region, the spinal canal, the lumbo-sacral nerve plexus, and several small venous branches from the lobes of the kidney. It passes dorsalward through the kidney and empties into the iliac vein. There are also communications with the vena intervertebralis thoracica.

The **vena renalis magna** (Fig. 64, No. 47) forms the main descending vein to the middle and posterior lobes of the kidney. It lies ventrally and mesially on the middle lobe and partly on the inner part of the posterior lobe. It sometimes has on each side two main trunks. The vein receives some small vessels out of the anterior kidney lobe, also other small veins from that part of the peritoneum which covers the kidney, from the rectum and finally small veins from the ureter.

VISCERAL VEINS OF THE POSTERIOR VENA CAVA

The **venæ testiculæ**, or the vena ovariana drain the blood from the testicles of the male and ovary of the female. The size of these veins change with the enlargement of the testes or of the ovary during reproductive activity.

The **venæ suprarenales revehentes** are short, thick trunks coming from the adrenal glands. The left empties into the left side of the posterior vena cava and the right into the right dorsal side. These receive veins from the testicles in the male and from the ovary in the female.

The **vena proventricularis inferior** drains the stomach wall. One branch of this vein enters the left side of the posterior vena cava; the other enters the vena proventricularis communis, which in turn empties into the trunk of the anterior vena cava sinistra.

The **venæ hepaticæ** consist of one large and several small veins from each lobe of the liver, and empty into the posterior vena cava.

Some small vessels come from the pericardium and the peritoneal covering of the liver, and pass in the mediastinum to the trunk of the posterior vena cava.

In the region of the vena portalis:

The liver receives almost all the blood from the stomach, the intestines, the pancreatic gland, the spleen, partly from the liver itself, and partly from the abdominal air-sacs. This blood enters into the liver through the vena portalis dextra, the vena portalis sinistra, and the vena portalis propria.

In both lobes of the liver these veins divide into numerous small branches and collect again into two large short trunks, the vena hepatica magna dextra (Fig. 70, No. 23), coming out of the right lobe of the liver and the vena hepatica magna sinistra (Fig. 70, No. 24) coming out of the left lobe. These two vessels empty inferiorly into the posterior vena cava.

The vena portalis dextra receives the blood from the vena mesenterica communis.

The **vena mesenterica communis** (Fig. 64, No. 36) receives the blood from the vena coccygo-mesenterica (Fig. 70, No. 28) which comes from the arcus hypogastricus and receives the vena hemorrhoidalis (Fig. 64, No. 37). This drains the cloaca and the bursa of Fabricius, and it also receives veins from the rectum and from the base of the cæca.

The **vena mesenterica anterior** (Fig. 71, No. 3) accompanies the anterior mesenteric artery, and collects the blood from numerous vessels from the small intestine.

The **vena pancreatico-duodenalis** (Fig. 64, No. 38) comes out of the duodenum and the pancreas, along the right side of the stomach and along both cæca.

The **vena proventriculo-lienalis** (Fig. 64, No. 39) comes out of the dorsal side of the proventriculus on the left side of the gizzard, passes along the hilus of the spleen, and takes up several splenic veins.

The **vena portalis dextra** (Fig. 64, No. 40; Fig. 71, No. 4) receives a vein near the base of the gall-bladder. This branch enters the right lobe of the liver and unites with the vena portalis sinistra.

The **vena portalis sinistra** (Fig. 71, No. 5) enters the left lobe of the liver and there forms a sinus. It receives vessels which come from the muscles of the gizzard, the inferior vena proventricularis, and from the proventricular wall.

The **vena portalis propria** receives small veins which come out of the walls of the abdominal air-sacs and from the fat of the abdominal walls.

The **vena umbilicalis** originates in the umbilical region and empties into the vena hepatica magna sinistra at a point where it comes out of the liver (Fig. 70, No. 29). This is the remains of an embryonal vein which collected all the blood of the yolk sac, passed on the left side of the large intestine to the body, took up the vena mesenterica and ended as the vena umphalo-mesenterica.

THE LYMPHATIC SYSTEM

The peculiarity of the lymph vessels is that they are associated with organs in which lymph cells are formed.

The lymphatic system consists of the lymph vessels and the cell-forming organs. In some instances the cell-producing organs are lymph follicles and in others lymph glands. For the most part the glands are replaced by plexuses which in many places surround the blood-vessels.

The lymph of birds is similar to that of mammals. The larger lymph vessels are similar to the veins, although the walls are always thinner. Its tunica intima is rich in elastic fibers and has a layer of endothelial cells on the inner side. The tunica media is formed of rings of smooth muscle fibers. The adventitia is composed of loose connective tissue.

The lymph vessels frequently form plexuses. The large lymph trunks follow the course of the larger blood-vessels, and frequently surround the arteries.

All the lymph vessels of the body, exclusive of the lymph of the caudal region, form into a large trunk which originates on both sides of the celiaca communis and passes upward along the side of the abdominal aorta, reaching a point anterior to the celiaca. By receiving many vessels in this region it forms a plexus around the aorta, and finally divides into two vessels, the right and the left, ducti thoracici.

The lymph vessels of the left side of the head, the neck, and the lung, and the left wing, and also lymph vessels of the proventriculus and the throat enter into the left ductus thoracicus. They accompany the jugular vein and are closely associated with the thyroid gland.

The right thoracic duct receives the lymph veins from the right cervical lymph vein, and the right side of the head, the neck, the lung, and from the right wing.

After the right cervical lymph vein has passed through the right thyroid gland, it divides into two branches, one branch emptying into the right thoracic duct and the other into the vena cava dextra.

The lymph vessels of the liver, the stomach, the pancreas, and the duodenum enter near the root of the arteria celiaca into the large lymph trunk. The lymph vessels of the remainder of the intestines, of the kidney, and of the generative organs empty farther caudally.

The lymph vessels of the intestines take up the emulsified fat. This emulsion in birds is colorless. The vessels pass upward along the mesenteric arteries. There are no mesenteric glands. These vessels form a plexus around the arteria celiaca. The lymph vessels of the posterior extremities accompany the artery, especially the anterior iliaca externa, and empty into the thoracic duct at the point of the anterior iliaco-communis.

The lymph vessels in birds are numerous. The lymph glands are few. They are only visibly found in the anterior breast and the neck region, and sometimes in the wings. Lymph follicles are numerous in the intestines.

The thin walls of the lacteals, of the lymph vessels, and of the thoracic duct are made up of two tunics, the inner being the thinner and weaker.

The lymphatics of the foot unite to form the vessels along the sides of each toe. In palmipedes there are anastomosing branches which pass from the lateral vessels of one toe to those of the adjoining toe, forming arches in the uniting web of the foot. These branches form a small plexus at the anterior part of the digito-metatarsal joint, from which pass three or four lymph vessels. The anterior and internal branches accompany and form a network around the blood-vessels. The posterior and external branches receive the lymphatic vessels from the sole of the foot. They then ascend along the metatarsus and form, at its proximal articulation, a close network from which vessels pass along the tibial region, forming a plexus around it as far as the middle of the leg. From this there arises two branches. The smaller passes along the anterior part of the depression between the tibia and the fibula, as far as the knee-joint, where it joins the other branch, which accompanies the blood-vessel. The trunk formed by the union of these two vessels accompanies the femoral vessels. Forming plexuses in its course, it receives tributary vessels from the adjacent muscles. The iliac trunk accompanies the femoral vein into the abdominal cavity, entering just in front of the anterior end of the pubis. At this point it receives branches from the lateral parts of the pelvis and then separates into two branches. The posterior vessel receives some lymph from the anterior lobes of the kidney, and from the ovary, or testis, and communicates anteriorly with a branch formed by the lymph vessels adjacent to the anterior mesenteric artery, and posteriorly with a large vesicular plexus surrounding the aorta and its branches. This plexus receives the lymph from the renal plexus and from those accompanying the arteria media.

There are two sacral or pelvic vesicles which are situated at the angle between the tail and the thigh in the posterior part of the abdominal cavity. Each vesicle is a trifle more than a half inch long and a quarter inch wide, and is shaped somewhat like a kidney bean. They have muscular coats with striated fibers. These sacs are called "lymph hearts."

The anterior division of the femoral lymphatic trunk accompanies the aorta, on which it forms a plexus with the branches of the opposite side, and with the intestinal lymph vessels. These vessels commence from a continuous plexiform network located between the mucous and the muscular coat of the intestine. They are larger at this point than where they leave the intestine to pass through the mesentery. They accompany the trunk of the anterior mesenteric artery and form a plexus around it.

Before reaching the region of the aorta, the intestinal lymphatic vessels communicate with the posterior division of the femoral trunk and with the lymph vessels of the ovary or of the testis. After passing to the region of the aorta they receive vessels from the pancreas and the duodenum, and terminate around the celiac axis with the lymphatics of the liver, the proventriculus, the gizzard, and the spleen, forming a rather voluminous plexus (Lauth).

The aortic plexus represents the receptaculum chyli and gives origin to two thoracic ducts, mentioned above, which passing on each side of the bodies of the vertebræ, pass one right and one left, over the lungs, from which they receive lymph vessels, and terminate after receiving the lymph vessels of the wing, into the jugular vein of their respective sides. The left thoracic duct, before emptying into the vein, receives the trunk of the lymphatics of the left side of the neck, and the right duct that of the right side of the neck, each tributary collecting lymph from all the structures of its side.

The lymphatics of the wing follow the course of the brachial artery, forming a plexus around it. These vessels are well developed in the elbow region. The principal trunk, following the humerus, receives collateral branches in its upper third. This vessel, when nearing the chest, receives two or three large lymph vessels from the pectoral muscles, and a branch which accompanies the brachial plexus.

The lymph vessels of the head accompany the branches of the jugular vein, collecting the lymph from the structures of the head and the neck.

The lymphatic vessels communicate at the anterior and posterior oblique anastomosing vessels. At the lower part of the neck each trunk receives a vessel, which accompanies the carotid arteries. Further on they are provided with a lymph gland which rests on the jugular vein.

THE BLOOD AND ITS FUNCTIONS

The special function of the blood is to nourish all the tissues of the body, and in this way to aid growth and repair. It furnishes material for the purpose of the elaboration of body secretions; it supplies the organism with oxygen; and it carries away carbon dioxid and other effete material. Blood is constantly in circulation.

Blood is red, opaque, and is, in the fowl, quite viscid. The exact tint of the blood depends on whether it is drawn from an artery or from a vein. Blood from a vein has a purplish tinge while that from an artery is a bright scarlet. The color of blood is largely due to pigment in the erythrocyte, called hemoglobin.

The *reaction of the blood* is alkaline, due to the phosphate and the bicarbonate of soda. The alkalinity of the blood is reduced by work. This is due to the formation of sarcolactic acid in the muscle. The odor of blood is due to volatile fatty acids. Each kind of fowl has its own peculiar odor. The taste of the blood is saltish, due to a small amount of sodium chlorid it contains.

The blood consists of the following substances:

First, the unorganized part, or fluid, the liquor *sanguinis* or *plasma*. It contains in solution proteids, extractives, mineral matter, and gases. The gases are held in loose chemical union.

The liquor sanguinis constitutes fully 66 per cent. of the volume of the blood. It is albuminous in nature and contains a small amount of coloring matter of a fatty nature. It holds in solution three proteids—fibrinogen, serum globulin, and serum albumin.

Second, the organized parts, or the cellular structure (Fig. 74, Nos. 1 to 21). The cells float in the plasma and consist of three groups: the erythrocytes, or red blood cells, the leucocytes, or white blood cells, and the thrombocytes.

Erythrocytes.—The average number of red cells (Fig. 74, No. 21) of the domestic fowl range between 3,000,000 and 4,000,000 per cubic millimeter. The red blood cells are flattened and elliptical in shape, and possess an oval elliptical nucleus. The average length is $1/2100$ inch and the diameter $1/3800$ inch, or 7 to 8 micra in diameter and 12 to 13 micra in length. However the diameters vary in different kinds of birds. The cytoplasm is yellow and glassy, and the nucleus takes basic stains and appears somewhat picnotic.

Thrombocytes.—The thrombocyte (Fig. 74, No. 19) is of about the same length as the erythrocyte but somewhat narrower. The nucleus is round, stains purple with the Wright's stain, and the chromatin material is somewhat diffused. The diameter of the nucleus is nearly equal to that of the cell. The cytoplasm is pale and may show vacuoles near the nucleus. They may contain small circumscribed red structures. They vary somewhat in size and shape. There are in the domestic fowl between 45,000 and 55,000 per cubic millimeter.

Leucocytes.—There are, in the blood of the hen, 28,000 to 35,000 leucocytes per cubic millimeter. The leucocytes may be divided into five distinct types. These are as follows:

Polymorphonuclear leucocytes with eosinophilic rods (Fig. 74, No. 2) are round and have a diameter about equal to the length of the erythrocyte. The nucleus is polymorphous; that is, it has two or more lobes. The nucleus stains a pale blue, and the chromatin is diffused. The cytoplasm is colorless with bright red staining spindle-shaped rods. There are 28 to 32 per cent. of this type of cell in the blood of the hen.

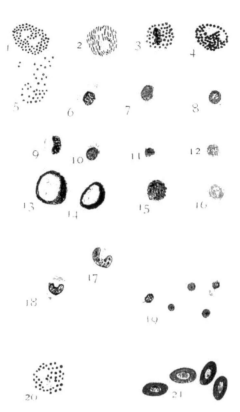

FIG. 74.—Blood cells of the fowl. *Wright's stain.* From a S. C. Rhode Island Red cockerel. 1, Basophile (mast cell) with lilac staining spherical granules. 2, Eosinophile with rod-shaped acidophile bodies. 3 and 4, Eosinophiles with acidophile staining round granules. 5, Eosinophile ruptured. 6, Mononuclear leucocyte with vacuoles in the cytoplasm. 7 and 9, Transitional leucocytes (first stage). 8, Mononuclear leucocyte. 10, 11, and 12, Small lymphocytes. 13 and 14, Large lymphocytes. 15, A lymphocyte. 16, Lymphocyte (nucleus centrally located). 17 and 18, Transitional leucocyte. 19, Thrombocyte. 20, Neutrophile (polymorphonuclear). 21, Erythrocytes.

Polymorphonuclear leucocytes with eosinophilic granules (Fig. 74, No. 3 and 4) are of about the same shape and size as the preceding. The nucleus is similar to the former except that it may appear slightly picnotic. The cytoplasm stains not at all or faintly blue; it contains round or spherical granules which stain a dull red. There is from 4 to 6 per cent. of this type of eosinophiles found in the domestic fowl.

Lymphocytes are round in shape and of about the diameter of the width of a thrombocyte (Fig. 74, No. 11). The nucleus is round, staining somewhat purple, and contains a diffused chromatin material. The cytoplasm exists in only small amounts; it lies to the side of the nucleus and stains a pale blue. This is the small lymphocyte. A similar cell but much larger also exists. This is the large lymphocyte (Fig. 74, No. 13). There is from 40 to 44 per cent. of the lymphocytes in the blood of the fowl. The small lymphocytes are most abundant.

Large mononuclear cells (Fig. 74, No. 6) either round or oval, in shape, whose diameter may be about that of an erythrocyte and at times much larger. The nucleus may be round, oval, or irregular, and at times rather crescent or U-shaped (Fig. 74, No. 17). The cytoplasm is abundant and completely surrounds the nucleus. The cytoplasm stains a paler blue than the nucleus. Both taking the basic stain as do the lymphocytes. These constitute 18 to 20 per cent. of the cells of the blood.

Mast cells or *basophiles* (Fig. 74, No. 1) are of about the same size and shape as the eosinophiles. The nucleus is round or oval, and stains a very pale blue. The cytoplasm is colorless, mostly to one side of the nucleus, and contains round or spherical purple staining granules. This type of cell constitutes from 2 to 4 per cent. of the white cells of the blood.

Structure of the Red Blood Cell.—The red blood cell is composed of a spongy stroma holding in its meshes the red coloring matter. The stroma, or framework, of the erythrocyte consists principally of nucleo-albumin; it contains lecithin, cholesterin, and salts. The red matter consists of an albuminous crystalline substance called hemoglobin, which forms about 90 per cent. of the total solid matter of the dried corpuscle. Each red cell offers a certain absorbing surface for oxygen. As the blood circulates through the delicate walls of the lungs and the air-sacs, it takes up oxygen; the blood at the same time delivers to the air carbon dioxid which has been brought from the tissues where active cell metabolism has been going on. This oxygen taken up by the erythrocyte forms a loose chemical union and is known as oxy-hemoglobin. In this form it is carried to the tissues of the body where it is given up by the erythrocyte to the tissues where oxidation is going on.

Hemoglobin is a crystallizable proteid substance containing carbon, hydrogen, oxygen, nitrogen, sulphur, and iron.

Formation of the Cells of the Blood.—The red blood cells are formed in the red marrow of the bone.

Polymorphonuclear leucocytes are formed in the red marrow of the bones, and the lymphocytes in the lymph glands and lymph follicles.

The bird carries a normal body temperature of 105° to 107° F. The average temperature of 50 mature hens and cocks was 106.8° F. The blood is of a deep red color.

Composition of the Blood.—The average composition of the blood of the domestic fowl as given by Owen, is as follows:

Whole shed blood:		
	Water	780 parts
	Clot	157 parts
	Albumin and salts	63 parts
		1000 parts
Moist blood cells:		
	Average total weight	456.69
	Water	342.52
	Solid matter	97.50
Plasma:		
	Total weight	543.30 parts
	Water	495.72 parts
	Solid matter	30.72 parts

Blood when drawn and allowed to stand soon coagulates. In the blood of birds this process is very rapid, the blood coagulating, in most instances, in about one-half minute. Blood coagulates only in the presence of calcium salts.

During life, the liquor sanguinis is termed plasma; but after it has been shed from the body and coagulation has taken place, the liquid residue is called serum. Serum is plasma with its modifications as the result of coagulation, and as this latter process is brought about by the production of fibrin, we may say that serum is plasma minus fibrin-forming elements.

The proteids of the serum are serum globulin, serum albumin, and a ferment produced as the result of coagulation. As fibrinogen is used up in the process of coagulation, it is not found in the serum, but there is in the serum a proteid known as fibrino-globulin. This is produced from fibrinogen during the process of fibrin formation. The following tabulation gives a clear idea of the difference between the proteids of plasma and of those of serum:

Proteids of Plasma

Fibrinogen

Serum globulin

Serum albumin

Proteids of Serum

Serum globulin

Serum albumin

Fibrin ferment (nucleoproteid)

Fibrino-globulin

Fibrinogen is the precursor of fibrin.

The fibrin of the blood clot of the bird is soft and very lacerable. The serum is usually yellow.

THE FATE OF THE ERYTHROCYTE OF THE FOWL

The power of vascular endothelium to ingest red blood corpuscles has been studied by Keys.

When bacteria or other minute foreign bodies are injected into the blood stream of pigeons, they are rapidly withdrawn from the circulation into the tissues of the liver and of the spleen. The foreign bodies are noted to be contained within cells of a distinct type, which is found in both liver and spleen. This type of cell contains, in addition to the foreign substances injected, much yellow pigment, and when tested for iron by Pearl's method gives a positive Prussian-blue reaction.

In such specimens there is a display of contrast to other tissues. There is an extensive content of cells possessing the distinct tone of Prussian-blue iron reaction. These cells are distributed rather evenly throughout both the spleen and the liver, but more numerously in the liver.

In the liver, under low-power magnification, these cells appear as blue patches, sharply differentiated from the red-stained parenchyma. These cells are larger in their greater diameter than the liver cells. They vary much in size and form. They bear a constant relationship to the venous capillaries, and often appear to occupy the lumen of the vessels. Under higher magnification it is noted, however, that each cell is an integral part of the endothelial intima lining of the capillaries. They are therefore fixed tissue cells, engaged by one of its surfaces upon the reticulum of the vessel wall, with a free surface bulging a greater or less degree into the lumen of the vessel. The attached surface of the cell follows exactly the line of the vessel wall. These cells are similar to those described for mammals by Kupffer and are called Kupffer cells or stellate cells. In the fowl Keys proposes the name hemophages. The nucleus of the hemophage stains a deep garnet with the carmine used in the above-given technic, and contains two or three very distinct and intensely stained nucleoli. In the hemophages, which are more nearly flat, the nucleus appears like those of the typical endothelial cells; whereas in the protruding hemophages of greater bulk, the nucleus is more vesicular and is irregularly pyramidal in form. Rarely two nuclei are found in one cell. Within this cell may be seen vacuoles of the cytoplasm which contain red blood corpuscles. These blood corpuscles have been phagocyted from the circulating blood stream. Approximately one-third of the intimal cells are hemophages. Each hemophage displays evidence that it contains, or has recently contained, one or more red blood cells. The cell body of the hemophage has no fixed morphology, but changes from time to time according to its phase of phagocytic activity. In a stage which the hemophage has recently ingested a red blood cell, the cell body bulges out into the lumen of the vessel and the nucleus is crowded to one side. At this time the red blood cell appears as those in the blood stream and possesses the characteristic staining reactions. The nucleus of the red blood cell stains deep reddish brown and the cytoplasm an even yellow bronze. In the next stage the cytoplasm of the hemophage gives a diffuse Prussian-blue reaction. Then in hemophages which represent later stages there are various stages of disintegration and digestion of the red blood cell. The first changes of the phagocyted red blood cell is hemolysis, the hemoglobin escaping into vacuoles of the cytoplasm of the phagocytic cell, leaving the nucleus-containing stroma distinctly outlined. The stroma may retain the original ovoid form or may become spherical; the nucleus in such instances remains ovoid. Gradually, both the stroma and nucleus lose their staining reaction until finally the vacuoles contract about a small indistinct remnant of the nucleus, which in its turn ultimately disappears. During this latter process the size of the hemophage gradually decreases. The hemoglobin, which has escaped into the cytoplasm of the hemophage, is seen to undergo a series of changes. At first the greater part of the pigment does not give the iron reaction but retains its yellow-bronze tone with erythrosin and occupies vacuoles of various sizes. Later the contents of the vacuoles give the iron reaction and with increasing intensity. Later there is a gradual decrease in the staining reaction indicating that the iron gradually disappears from the cells which extracted it from the red blood cells it digests. As a summary we find, that these cells take care of the worn out red blood cells. They devour them; hemolyze them, destroying the stroma and nucleus; split the hemoglobin and free the iron; and then finally return to their normal form.

The spleen contains the same type cells, but they are fewer in number. For the most part they are confined to the pulp cords and have no such evident relation to the vessel wall, or lumen, as in the liver.

The function of the cells of the spleen are essentially the same as those in the liver.

Iron freed from the worn out red blood cells is not retained by the cells freeing it, nor is it found in the bile. It does not occur elsewhere in the tissues of the liver and spleen. It is possibly discharged into the blood stream, and transported to the hemapoietic tissues. Cells which hemolyze red blood cells and liberate the iron are to be seriously thought of in connection with bile formation since bilirubin is approximately, if not identical with, iron-free hematoidon.

NEUROLOGY

The Nervous System.—The nervous system is an apparatus by means of which animals appreciate and become influenced by impressions from the outer world. Animals react on these impressions, and thus are enabled to adapt themselves to their environment. This system is the organic substratum of life, sensation, and motion. Broadly stated, the nervous system connects the various parts of the body with each other, and to coördinate the parts into a harmonious whole in order to carry on the bodily functions methodically and to control the physiological division of labor throughout the organism.

The nervous system consists of two parts. The first is the *cerebro-spinal system*, which comprises the central nervous axis, including the brain and the spinal cord, and the peripheral nerves, including the cranial and the spinal nerves. The second is the *sympathetic nervous system*. The two parts of the system are closely linked together, and both terminate in peripheral nerve endings, including those of special sense, of sensation, and of motion.

The cerebro-spinal nerves especially preside over the special senses, motion and sensation; and the sympathetic over the digestive, the pulmonary, and the vascular apparatus.

From a structural standpoint, the nerve system consists of cell elements peculiarly differentiated from all other tissue cells in that their protoplasm is extended in the form of processes, often to great distances from the nuclear region. The cell elements are held in place by supporting tissue and receive an abundant blood supply; they are partly of ectodermal and partly of mesodermal origin.

The cell element of the nerve system, called a *neurone*, is the developmental, structural, and functional unit of the nervous system. It is a single cell presenting unusual structural modifications. It comprises not only the nerve cell body with its numerous protoplasmic processes, or dendrites, but also the axone, which may vary in length from a fraction of a millimeter to fully half the bird's length. The bulk of the axone is many times the bulk of the cell body.

Certain non-medullated axones are surrounded by a delicate, homogeneous, nucleated sheath, called the neurilemma, or sheath of Schwann.

THE CRANIAL NERVES

The cranial nerves have their origin in the brain and leave the cranial cavity in pairs. They are numbered numerically from before backward, there being twelve pairs in all. The following is a tabulation according to their number, name, and function:

No.	Name	Functional Nature
I.	Olfactory	Smell-sense
II.	Optic	Visual-sense
III.	Oculomotor	Motor to muscles of eyeball and orbit
IV.	Pathetici	Motor to superior oblique muscle of eyeball
V.	Trifacial	Mixed: Sensor to face and tongue. Motor to face
VI.	Abducentes	Motor to External rectus of eyeball
VII.	Facial	Motor to muscles of head and face
VIII.	Auditory	Hearing-sense
IX.	Glosso-pharyngeal	Mixed: Tongue, pharynx and muscles of throat
X.	Vagus	Mixed: Sensori-motor to respiratory tract and part of alimentary tract
XI.	Spinal accessory	Motor to muscles of pharynx, neck and heart
XII.	Hypoglossal	Motor to muscles of the tongue

Olfactorius.—Nervus olfactorius (Fig. 75, *C*, 16). This is the nerve of smell, one of the nerves of special sense. The organ of smell consists of five layers as follows:

First, a layer of olfactory fibers extending in different directions and consisting of a dense plexiform arrangement of the axones of the olfactory cells. From this layer the fibers pass into the layer of olfactory glomeruli where their terminal ramifications mingle with the dendritic terminals of cells lying in the more dorsal layers, to form distinctly outlined spheroidal or oval nerve fiber nests, the *olfactory glomeruli*.

Second, a fine granular layer of basic substance containing round cellular structures, the *stratum granulosum*.

Third, broader granular, or molecular layer, having on its inner surface a row of large pyramidal cells which are both small and large and which send their dendrites into the olfactory glomeruli. Their points are directed outward.

Fourth, a layer of round cells tightly pressed together and measuring about 5 microns in diameter. Between these cells are very fine nerve fibers.

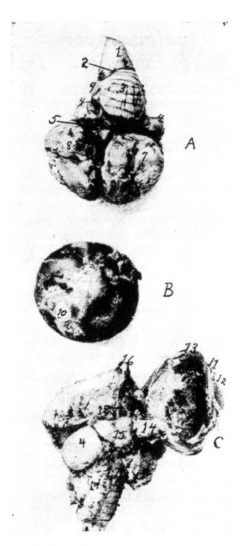

FIG. 75.—The brain of a hen. Photograph.

A. Upper surface of the brain. 1, Medulla oblongata. 2, Calamus scriptorius. 3, Cerebellum. 4, Optic lobes. 5, Transverse fissure. 6, Longitudinal fissure. 7, Upper surface of the left cerebral lobe. 8, Upper surface of the right cerebral lobe. 9, Lateral pillar of the cerebellum.

B. The posterior surface of the eyeball. 10, The sectioned surface of the optic nerve.

C. The inferior surface of the brain. 11, Corneo-scleral juncture. 12, The cornea. 13, The sclera. 14, The optic nerve. 15, The optic chiasm. 16, The olfactory lobes. 17, The medulla oblongata. 18, Tuber cinererum et infundibulum.

Fifth, a layer of epithelial cells.

The peripheral fibers and the nerve cell layers near the hemispheres disappear so that the basic substance of the trabecula with the hemispheres form the entire lobe mass. On the lower surface of the hemispheres there is a long bundle of nerve fibers which enter into the substance of the olfactory lobes and there disappears. These fibers are medullated. Non-medullated fibers enter into the makeup of the olfactory trunk.

The original center of the olfactory nerve is not in the hemisphere, but in the same location as the optic nerve. This nerve trunk consists of very fine non-medullated fibers. The nerves of smell are therefore not peripheralistic nerves. The nerve fibers are distributed to the mucous surface of the turbinated bones. Toward the front they form an expanded prolongation.

The olfactory nerve, as it emerges from the cone-like anterior tip of the cerebrum, is of considerable thickness and extends along with the median dorsal artery of that region above and to the inside of the orbit, under the thin bony structure. Before its termination into the posterior turbinated bones, it is crossed by the superior maxillary division of the fifth pair of cranial nerves. It extends as far as the pituitary membrane of the turbinate bones upon which its filaments are distributed radially.

The Opticus.—The nervus opticus (Fig. 75, No. C, 14) the second cranial, is the nerve of sight. The optic lobe, or tuberculum bigeminum lies at the base of the brain on each side of the optic tract.

From the optic lobes the two trunks pass forward along the under surface of the cerebri, forming the optic chiasm at the hypophysis (Fig. 75, No. C, 15), at which point the nerve fibers originating from the right side pass to the left, and *vice versă*. From the chiasm the true optic nerves pass forward to the posterior surface of the eyeball. They are composed of a bundle of very fine, marrow-like nerves. These fibers enter into the ganglionic cells of the retina. By removal of the finely adherent neurilemma, the optic nerve is seen to be composed of parallel, longitudinal lamellæ, the margins of which are mostly free on one side.

The Motor Oculi.—The motor oculi (Fig. 62, No. C, 10), the third cranial nerve is a motor nerve. It originates close to the base of the brain behind the position of the hippocampus of mammals on the inner side of the crus cerebri, and also on the inner somatic column close below and somewhat aside from the aqueduct of Sylvius. The nerve leaves the brain cavity through a distinct foramen near the foramen opticum, and supplies the following eye muscles: first, after entering the orbit, it sends a branch upward to the inferior portion of the inferior rectus; then, it gives off the thick ramus ciliaris. The trunk then extends under the optic nerve and passes forward to innervate the inferior rectus, the internal rectus, and the inferior oblique. A ciliary ganglion forms on the trunk.

The Patheticus.—(Fig. 62, No. *C*, 11) is a small motor nerve, originating close to the sulcus centralis, in the circle of the center brain over the valve of Vieussens, between the posteriors of the optic lobes. It extends in a dorsal direction between the cerebellum and the lobus opticus, to the posterior of the latter of which it then forms a loop ventrally. Lying close to the optic foramen it passes through a fine opening into the eye cavity and supplies the superior oblique muscle of the eye. During its course it passes dorsally over the optic nerve, and then crosses dorsally over the ophthalmic division of the fifth pair of cranial nerves and the internal rectus muscle.

The Trifacialis.—(Fig. 62, No. *C*, 12). This, the fifth cranial, is a mixed nerve and is divided into two parts, portio major and portio minor.

The *portio major* originates in the ganglion cells of the posterior part of the brain. Commencing near the medulla oblongata it passes through the posterior part of the brain, then upward and outward; and along its route it forms the Gasserian ganglion, which lies partly in the cranial cavity or in its wall.

The *portio minor* consists of the downward passing fibers containing the motoric elements which are distributed to the muscles of the jaw and of the eye. The roots are found in the ganglion from the center and back brain, close below the pathetic nerve origin, where the pathetic passes between the lobus opticus and the pars peduncularis; it then takes a lateral course between the two.

The fibers of the portio minor do not take part in the formation of the Gasserian ganglion, but are only partly surrounded by it.

There are three nerve trunks given off of the trifacialis: the ophthalmic, the superior maxillary, and the inferior maxillary.

The *ophthalmic division* of the fifth nerve is the smallest of the three branches. It emerges directly from the Gasserian ganglion, and passes through a narrow bony canal in the base of the brain, below the pathetic and the abducens, and through the foramen ophthalmicum. It then enters the eye cavity above the optic nerve, and on the wall of the cavity extends downward, and with regard to the eyeball, dorsally lying close to the rectus internus muscle and close to the surface of the eyeball. It passes along the olfactory nerve, extends under the superior oblique muscle, and finally reaches the inner angle, or canthus, of the eye. It here divides into the recurrent externa and the ethmoidalis. The ethmoidalis branch is a straight continuation of the ophthalmic. It extends along close to its fellow of the opposite side, and, over the vomer, splits into two branches. The smaller of these branches breaks through the bone cells of the jaw, continues upon its ventral surface in a furrow extending forward, and terminates in the gum region. It supplies the gum and point of the beak. The larger branch enters into the cell substance extending to the tip of the beak, in its course sending out a number of fine filaments, which spread out along the outer surface. At this point these two branches may fuse. The larger branch is endowed with the sense of touch.

The recurrent externa divides shortly after leaving the main trunk of the eye cavity passes over the lacrimal gland. It gives two branches to this gland and one to the membrana nictitans, and innervates the upper eyelid. It then emerges from the eye cavity, passing over the os lacrimale, and gives one or more branches to the integument of this region, including the comb. This branch is large in birds with a large comb. Branches pass in front of the lacrimal bone through the outer nasal cavity and into its deeper structure.

Shortly after the ophthalmic branch has entered into the eye cavity and before it crosses the optic nerve, it gives a fine branch to the motor occuli.

The second and the third branches of the fifth cranial nerve are mixed. They contain elements of both the portio major and the portio minor. These two branches come from the lower part of the outer region of the ganglion, and pass together through a cavity which is located between the os petrosum and alæ and basis sphenoid, and then branch.

The ramus secundus, or *superior maxillary division*, of the trifacial, is the second branch and passes into the orbit below the optic nerve and the eyeball. It is called the recurrent infra-orbitale. It gives an ascending branch to the gland of Harder, one to the conjunctiva, one to the membrana nictitans, and one to the eyelids. It also gives a branch to the skin below the eye and to the angle of the mouth. This latter branch is called the *recurrent subcutaneous*. These two branches communicate with the recurrent nasal ciliaris of the first trigeminus branch. The second branch passes below the nasal opening, then passes on and forms the alveolar nerve on the side between the gums. It sends several recurrent posterior branches to the elevations on the back part of the mouth. It extends forward to the point of the beak.

The *inferior maxillary division* of the fifth is larger than the other two. It is directed downward and outward then upward to the temporal region. At the temporal region it divides into five parts. It gives a branch to the temporalis muscle, one to the pterygoid muscle, and one to the mylo-hyoideus. The main branch gives twigs to the parotid gland region and enters into the dental canal of the lower jaw. Numerous filaments break through the lower jaw bone, and then spread out on the skin and rim of that bone. The largest branch, called the recurrent maxillaris externa, comes out near the coronoid process. The rest of the trunk extends to, and comes out of, several foramina at the anterior point of the jaw bone.

These filaments are contained in grooves or cavities of the lower jaw and terminate in touch buds.

The trifacialis fuses with the other cranial nerves and with the sympathetic nervous system. Some of these fusions are as follows: The recurrent ophthalmica fuses with the orbito-nasalis ganglion. Indirect fusion of the recurrent maxillary division of this nerve takes place near the Gasserian ganglion through the sympathetic nerves, the temporo-lacrimalis, the facial nerves, the large cervical ganglion, and indirectly with the glosso-pharyngeal and the vagus. This fusion has been called the superior recurrent branch of the trigeminus, or trifacialis.

There is a direct fusion of the superior recurrent maxillary division of the fifth just before entering the upper jaw. This fusion is with the spheno-palatine ganglion and the sympathetic carotidis cephalica, and also with the large cervical ganglion.

The Abducens (Fig. 62, No. *C*, 13).—This nerve originates in the somatic and motoric column along with the other nerves of the eye muscles. Its nucleus lies in the circle of the pars commissuralis of the cerebellum. The nerve then passes ventrally, as does the motoris oculi and leaves the brain along the median line. This is a comparatively large nerve and passes somewhat laterally and ventrally from the foramen opticum through a canal in the sphenoid. It then enters the orbital cavity. Some muscular twigs are given off to the quadratus and to the pyramidalis muscle. This nerve also innervates the external rectus muscle of the eye. The abducens anastomoses with the ramus ciliaris and the ophthalmic nerve and gives fine nerve twigs to the ramus ciliaris externus of the ciliary ganglion.

The Facialis and the Acusticus.—The facial nerve originates with the auditory (Fig. 62, No. *C*, 14 and 15) in a very vaguely known manner, from the cerebellum. It divides into three parts.

The first of these probably comes from the complex ganglion with the posterior roots of the auditory. This part belongs to the somatic sensory group of nerves. From this same group originates the auditory, which spreads out into the cochlea and takes the impressions of sound. This nerve is short and thick, and, at the point where it loses its medullary covering on entering the cochlea, there is developed a ganglion. This ganglion is similar to the spinal ganglia.

The second part is provided with one root which originates mesially and ventrally, from the deeper ganglion cells. Some of the fibers from this root constitute the vestibular branches. They accompany the auditory and supply the anterior part of the ear labyrinth and the semicircular canals. The larger part of the fibers of this trunk make up the intermediate part of the facial. The geniculate ganglion is formed at their fusion. The sympathetic spheno-palatine nerve emerges from this ganglion, coming out of the aqueduct of Fallopius.

The third part, called the portio dura, is the main facialis. It is located opposite the auditoria intermedia. Its roots may be traced to the complex ganglion, from which they take a direction ventrally from the median portion.

The facialis, after emerging from the aqueduct of Fallopius, takes a curved course and partly fuses with the sympathetic temporo-lacrimalis on the upper posterior wall of the ear drum. It is here accompanied by the carotid and the cephalic arteries. It leaves the ear drum through an opening in the quadrate bone, giving off a large branch to the digastricus muscle and a small one to the stapedius of the columella auris. The facialis trunk is quite large. It then passes downward along the quadrate bone, where it receives a large branch from the glosso-pharyngeus. This nerve also gives branches to the mylo-hyoideus and the stylo-hyoideus muscle, crosses laterally the glosso-pharyngeus, and finally fuses with branches of the subcutaneous and with the first, the second, the third, and the fourth cervical nerves. After this fusion the nerves innervate the skin of the anterior of the neck and the constrictor colli muscle.

The facialis anastomoses as follows: with a fine branch of the sympathetic temporo-lacrimalis, indirectly with the ramus trigeminus, with the large cervical nerve ganglion and with the spheno-palatine ganglion and nerve.

The Vagus Group.—The ninth, the tenth, and the eleventh cranial nerves are by some anatomists called the vagus group, first so called by Willis. They are made up of both sensory and motor nerves.

The Glosso-pharyngeus.—The roots of the glosso-pharyngeus (Fig. 62, No. *C*, 16; Fig. 76, No. 13) emerge, along with those of the vagus, from the medulla oblongata, and enter, as a short trunk, the foramen jugulare et caroticum. Between these two nerves there is usually found a thin portion of bone. The trunk of the glosso-pharyngeus forms a ganglion in the foramen jugulare et caroticum where it receives connecting branches from the near-by ganglion radicis and vagus nerve. The glosso-pharyngeus passes out of the foramen jugulare et caroticum above the large superior sympathetic nerve ganglion, with which it communicates. The glosso-pharyngeus passes diagonally over the ramus temporo-lacrimalis of the sympathetic system, and then receives a very strong branch from the vagus (Fig. 76, No. 12). It sends a short branch to the recurrent lacrimalis of the sympathetic system. At this point there is formed a reddish-yellow ganglion, the petrosal ganglion, which is similar to the petrosal ganglion of mammals. Frequently it is found close below the large superior cervical nerve ganglion. The petrosal ganglion is frequently connected by special fibers with the cervical nerve ganglion, the large superior cervical nerve ganglion and the ganglion radicis vagi. The glosso-pharyngeus is divided into the following branches.

First, the recurrent pharyngeus which gives branches to the upper part of the throat and which is tortuous in its course. It receives a branch from the superior cervical nerve ganglion and gives off branches to the salivary glands and papillæ of the posterior tongue region.

Second, the recurrent lingualis, which passes with the lingual artery over the hyoid bone to the base of the tongue and to the papillæ. Another branch passes below the tongue bone and supplies the tongue and the pharynx. The glosso-pharyngeal is a mixed nerve.

FIG. 76.—Photograph of spinal cord of a hen. 1, Cerebrum. 2, The olfactory nerve terminating at 3, the turbinated bones. 4, The superior maxillary division of the fifth nerve dividing. 5, The eye with the upper part laid open. 6, The cervical segment. 7, The lumbo-sacral segment. 8, The lungs. 9, The kidneys lying above the peritoneum. 10, The terminal filament of the spinal cord. 11, The hypoglossal nerve. 12, The vagus nerve. 13, The glosso-pharyngeal nerve. 14, The branch from the vagus.

The Vagus.—The vagus, or pneumogastricus (Fig. 18, No. 5; Fig. 62, No. C, 17; Fig. 76, No. 12), a mixed nerve, forms a ganglion in the foramen through which it passes. The foramen lies between the os petrosum and os occipitale, close to the foramen jugulare, above and to the inside of it.

The ganglionic root of the vagus communicates with the superior cervical nerve ganglion. The vagus, after coming out of the foramen, takes up branches of the spinal accessory, passes to the superior nerve ganglion, crosses the carotid artery, and then, accompanied by the internal jugular, it connects by a branch with the hypoglossal. It then communicates with the petrosal ganglion, and receives a long branch from the superior cervical ganglion. It then extends along the neck with the jugular vein and often fuses with the sympathetic ganglionic plexuses. As it passes, it is interwoven with the glosso-pharyngeus, but each nerve element retains its own individuality. Extending down the neck, on entering the thoracic cavity, it lies between the plexus brachialis and the carotid artery. Then it passes below the subclavian artery, and between the bronchial tubes, the aorta pulmonalis, and the subclavian vein. Then ventrally it rests upon the glands, and the right and the left vagi fuse or unite. From here they radiate down to the stomach in fan-shape, and, continuing, they fuse with the sympathetic system.

The other branches of the vagus are:

The first branch is the recurrent laryngeus, which supplies the lower end of the bronchial tubes, and the esophagus, then enters above the bronchus near the origin of the subclavian artery as the recurrent cardiacus.

The second are the recurrent pulmonale which pass into the lungs, each fusing with its fellow from the opposite side, and giving off branches inferior to the vena cava including a branch to the heart.

The third are the recurrent hepatici which pass through the diaphragm and are distributed to the liver.

The Accessorius Spinalis.—This is a very small motor nerve (Fig. 62, No. C, 18). It comes out between the dorsal and the ventral root of the third cervical nerve. It lies close to the neck and extends anteriorly, receiving roots from the first and the second cervical nerve. It passes through the occipital foramen into the brain cavity and then enters the ganglion radicis vagi. It passes out of the cranial cavity through the foramen jugulare. It then partly fuses with the vagus and partly, as a fine branch, with the subcutaneous colli.

The Hypoglossus.—The hypoglossus is a motor nerve (Fig. 62, No. C, 19). It originates anterior to the eleventh pair of cranial nerves, and from the same ganglion as the abducens and the motor oculi. It leaves the medulla oblongata from its ventral surface, and passes with a posterior and an anterior branch out of the cranial cavity through two separate foramina in the os occipitale laterale. The posterior branch, much smaller than the anterior, passes between the cranial bones and the cervical sympathetic nerve; it then passes at right angles through the rectus capitis anticus, and, while closely following the carotid artery, it fuses with the anterior branch. The larger, anterior branch, much stronger than the posterior, sends a short strong branch to the complexus muscle. The rest of this branch, which probably has sympathetic elements, is disposed as is a spinal nerve. It crosses the sympathetic nerve, and at this point forms a typical cervical sympathetic ganglion. It also forms a short loop with the recurrent ventralis of the first cervical plexus. From this loop, after it has given off several branches to the muscles of the neck, it gives off one or two strong branches, which fuse with each other and with the posterior thin branch, thus forming the trunk of the hypoglossus. It receives many elements from the first cervical nerves.

The hypoglossus sometimes communicates with the vagus, crosses over the latter, and divides itself into two main branches, as follows:

First, the *recurrent laryngo-lingualis*, which passes between the cornua of the os hyoideum and the larynx to the anterior part, and furnishes the principal tongue muscles. It extends along the inferior surface of the tongue and fuses with the one of the other side and extends to the free tip of the tongue. This nerve probably receives sensory elements from the second root and from the confluent of the first cervical nerve. This form of anastomosis is well marked in birds with thick tongues, as ducks.

Second, the *recurrent laryngeus* furnishes the muscles of the superior larynx and of the tongue skeleton. It extends downward and also furnishes the muscles of the trachea. It follows the course of the jugular vein and at the entrance of the chest supplies the furcula. The nerve passes into the thorax, downward along the side of the bronchial tubes, and supplies innervation to all the muscles of the inferior larynx.

THE SPINAL CORD

Structure of the Cord.—The spinal cord is called the myelon or medulla spinalis. It is a comparatively large, white, irregularly cylindrical cord, flattened from above downward. It extends from the foramen magnum at the base of the medulla to the caudal portion of the spinal canal where it terminates in a fine filament. In order to allow considerable motion of the spinal column without danger of injury to the spinal cord, the cord is loosely suspended in the canal. The meninges, or coverings, are continued from the brain. In addition to these there are to be found arteries, veins, and nerves entering and others making their exit from the canal and from the cord. At the cervico-dorsal juncture, where the brachial plexus is given off, the cord is enlarged. There is another enlargement at the point where the lumbo-sacral plexus is given off. At this point, superiorly, there is a longitudinal cavity called the *sinus rhomboidalis*. This sinus contains a gelatinous substance. There is a pair of nerves given off at each intervertebral foramen. Thus there are as many pairs of spinal nerves as there are vertebral segments. The inferior root is the motor root and carries the impulses from the cord to the periphery. The superior root is the sensory branch and carries the impulses from the periphery to the cord, and thence the impulse is carried to the centers in the brain. The sensory and the motor roots are of about equal size. The inferior have more numerous filaments. The ganglion on the superior or sensory root is relatively large. In the sacral region the sensory and the motor branches pass through their own bony canal.

White *rami communicantes* are given off at each intervertebral foramen to the sympathetic system and gray rami are received from the sympathetic ganglia. Thus there is established a direct communication between the sympathetic and the spinal system. There is a ganglion on the superior root just outside the spinal cord.

The ganglionic portion, or *gray matter*, is arranged in the center of the cord in two comma-shaped parts forming an X. The white matter, or the material forming the fibers, is arranged around the central ganglionic portion.

The spinal cord may be divided into two lateral symmetrical halves. There are two longitudinal fissures, one on the upper and the other on the lower half. The upper called the *superior median fissure* is narrow but deep. The one on the inferior side, called the *inferior median*, is usually more pronounced. The superior parts of the gray matter are called the superior cornua, or horns; the inferior parts the inferior cornua, or horns. The center of the cord is pierced by a *central canal* which communicates with the fourth ventricle at the calamus scriptorius.

STRUCTURE OF THE NERVE TRUNKS AND GANGLIA

The protoplasmic processes called *dendrites* have a similar structure to the cell body. The dendrites branch dichotomously, become rapidly smaller, and usually end at no great distance from the cell body.

The *axone* or *axis cylinder process*, differs from the cell body and from the dendrites. It does not contain chromophilic granules. It consists entirely of neurofibrils and perifibrillar substance. It emerges from the cell at an enlargement known as the axone hill. This hill is free from chromophilic bodies.

Nerves are divided into two kinds, medullated and non-medullated.

Medullated nerves are divided into two kinds: medullated nerves with a neurolemma, and medullated nerves without a neurolemma.

Medullated nerves with a neurolemma consists of an axone, a medullary sheath, and a neurolemma. A delicate membrane called an axolemma, or periaxial sheath envelops the axone. The medullary sheath called myelin, is semifluid, somewhat resembling fat. The outer covering is the neurolemma, or sheath of Schwann. It is a delicate, structureless membrane which incloses the myelin. There is under this sheath an occasional oval nucleus. At intervals there are constrictions, or nodes, called the constrictions of Ranvier. The part between the nodes is called the internode.

The medullated axones without a neurolemma are the medullated nerve fibers of the central nervous system.

The non-medullated axones, or *non-medullated nerve fibers*, are divided into non-medullated axones with and those without, a neurolemma.

The non-medullated axone without a neurolemma is merely a naked axone. They are confined to the gray matter and to the beginnings and endings of sheath axones, all the latter being uncovered for a short distance after leaving the nerve cell body and also just before reaching their terminations.

The long axones serve to make connections with the peripheral, or distant, nerve cell, muscle cell, or gland cell; while the shorter axones of certain neurones divide into terminal branches in the immediate vicinity of its cell body, presumably to come into relation with other nerve cells in the same or adjacent groups.

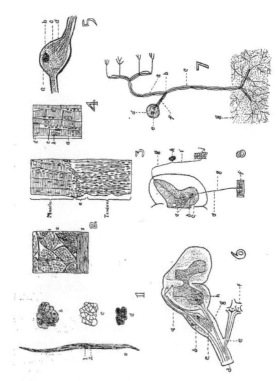

FIG. 77.—Histological structure of tissues.

1. Non-striated or involuntary muscle cell. *a.* 1, Nucleus; 2, Protoplasm. *b.* Transverse section showing nuclei in the center of the cells that are cut through the nuclear zone. *c.* A cross-section of the connective tissue which binds together the muscle cells. *d.* A section showing the so-called intercellular bridges.

2. A cross-section of the rectus abdominalis. 1, The endomysium which binds the cells into primary fasciculi or bundles. 2, The perimysium which surround the bundles of fasciculi. 3, The epimysium which surrounds the muscle. These binding structures are white fibrous connective tissue.

3. A longitudinal section at the juncture of a muscle and tendon. *a,* The juncture between the muscle and tendon. Note the many nuclei in both muscle and tendon. This is voluntary or striated muscle which make up the dermal, dermo-osseous and skeletal muscles.

4. A longitudinal section of heart muscle. *a,* Connective-tissue cell. *b,* Nucleus of a muscle cell. *c,* Cement line between the muscle discs. *d,* The cell or segment.

5. A bipolar ganglionic nerve cell. *a,* The nucleus. *b,* The nucleolus. *c,* The fibrillar structure. *d,* The medullary sheath.

6. A diagram showing the scheme of the peripheral nerve trunk. *a,* The neuraxis of the peripheral sensory neurone. *b,* The spinal ganglion of the superior or sensory root. *c,* The dendrite or peripheral nerve fiber of the sensory nerve. *d,* The nerve trunk. *f,* The sympathetic nerve ganglion connected with the spinal cord through the white and the gray ramus communicans. *e,* The neuraxis of the sympathetic neurone. *g,* Neuraxis or trunk of the motor neurone or nerve cell. *h,* The anterior horn of the gray matter of the spinal cord.

7. A diagram of a peripheral sensory neurone. *a,* The neuraxis which ends in the spinal cord or brain. *b,* The T-shaped division of Ranvier. *c,* The dendrite or sensory nerve fiber in the nerve trunk. *d,* The nucleus and nucleolus of the cell *e., f.* the axis cylinder process of the cell. *g,* The telodendrions or terminal branches of the dendrite or axis cylinder.

8. A schematic diagram of the sensory motor reflex. *a,* The telodendria. *b,* The dendrite. *c,* Nerve cell of the motor neurone. *d,* The motor neurone. *f,* The muscle fiber. *g,* The neuraxis of both sensory and motor neurones, the upper being the sensory. *h,* The nerve cell in the sensory ganglion. *i,* The

sensory neurone or axis cylinder (nerve fiber). *j*, The skin with peripheral telodendrion of sensory neurone.

Neurones are devoted to the maintenance of functions. Reproductive neurones are so arranged as to receive afferent nerve impulses from other tissues; emissive neurones give off efferent nerve impulses. The former are sensory neurones; the latter are motor neurones if connected with muscles, and excito-glandular if connected with gland cells. The basis, then, of the nerve system is a series of neurones, with projecting and association processes, coördinated for the purpose of performing specific actions manifested either by motion, by trophic changes, or by the apperception of stimuli of a chemic, mechanic (tactile and auditory), thermal, or photic nature.

The whole of the nerve structure is composed of the nerve tissue and supporting connective tissue. The neurones constitute the nerve tissue, while the supportive tissue is composed of neuroglia and of white fibrous tissue derived either from the investing membrane or from the sheaths of its numerous vascular channels.

The neurones, or nerve cells, exhibit marked variations as to external characters, dimensions, and form. The neurone presents a swollen cell mass and a nucleus; it is known as the ganglion cell. From this cell body are given off a number of processes of two distinct kinds: first, protoplasmic processes, which are commonly branched, called dendrites; second, a single, thinner, and paler process; the axis cylinder process.

The bodies of cells vary in size from 4 to 100 microns or more in diameter. The largest cells occur in the inferior horn of the spinal cord, in the spinal ganglia, in the large pyramidal cell layer of the cerebral cortex, in the Purkinjean cell layer of the cerebellum, and in Clark's column of the spinal cord. Very small cells occur in the olfactory bulbs, in the granular layer of the cerebral and cerebellar cortex, and in the gliosum cornuale of the cord.

Nerve cells are classified according to the number of processes arising from the cell body, and neurones are referred to as unipolar, bipolar, and multipolar.

Unipolar cells are met with frequently in early stages of embryonic development, but are rare in adults, occurring only in the retina, in the olfactory bulb, and within the baskets of the Purkinjean cells of the cerebellum. The cells of the cerebro-spinal ganglia, except the cochlear and vestibular, are apparently unipolar, but they are developmentally and functionally of bipolar nature.

Bipolar cells are found almost exclusively in the peripheral sensory system, as in the olfactory membrane, in the retina, in the cochlear and vestibular ganglia, and in the cerebro-spinal ganglia of the embryo.

Multipolar cells are the most numerous and form the principal elements of nerve centers throughout the system. They are termed multipolar because of the greater or lesser number of dendrites given off in addition to the single axone.

The body of the nerve cell consists of a mass of protoplasm surrounding a nucleus. The cytoplasm of the nerve cell consists of two distinct substances: first, neurofibrils; second, perifibrillar substance. In most nerve cells there is a third substance called chromophilic bodies.

The *neurofibrils*, extremely delicate, are continuous throughout the cell body and all of its processes. Within the body of the cell they cross and interlace and probably anastomose.

The *perifibrillar substance* is a fluid or a semifluid substance which both in the cell body and in the processes surrounds and separates the neurofibrils.

The *chromophilic bodies* are granules or groups of granules which occur in the cytoplasm of all the larger and of many of the smaller nerve cells.

THE SPINAL NERVES

The first pair of spinal nerves come out between the atlas and the occipital bone. Each nerve divides into three branches. The first is the *anterior branch* which innervates the dorsalis, the biventer cervicis, the cervicales, and the caput posticus muscles. The second is the *recurrent ventralis* which passes mesially and downward from the recurrent communicans. It innervates the rectus capitis anticus muscle and joins a branch of the hypoglossal nerve. There is given off a white ramus communicans to the sympathetic nerve.

The second spinal nerve emerges from the opening between the first and the second cervical vertebræ and is similarly disposed as the first cervical. It gives off a branch called the recurrent ventralis which fuses with branches of the first cervical and the hypoglossal nerve. This nerve innervates the complexus muscle.

The succeeding cervical nerves emerge in a similar manner down the neck and are distributed to the muscles and other structures of the cervical region. The second, the third, and the fourth cervical nerve gives off branches which form anastomosing loops with the facial nerve.

According to Gadow the last cervical nerve passes between the last cervical and the first dorsal segment. There are thus fifteen pair of cervical nerves.

The first pair of *dorsal nerves* pass out between the first two dorsal vertebræ. The first few dorsal nerve branches innervate only the trunk muscles. Other branches supply the skin and the other adjacent integument. The ventral branches of the cervical nerves often communicate with the ventral branches of the first dorsal nerves. These branches are larger than the superior branches and aid in the formation of the brachial plexus. Smaller inferior branches are distributed to the scalenus and other muscles, and extend as far posterior as the intestines.

The dorsal branches of the spinal nerves in the lumbo-sacral region are very small, on account of the lack of extensive development of the muscles of this region. The elements entering into these nerve trunks are largely vasomotor nerves. In addition to the upper twigs supplying the skin and the other integument of the region, other branches descend into the abdominal cavity.

The caudal spinal nerves are not well developed and disappear in the region of the caudal vertebræ. The dorsal branches innervate the levator muscles, and the ventral branches the depressor muscles.

There is thus given off, throughout the vertebral column, a pair of nerves between each two vertebral segments, making as many pairs of nerves for the region as there are vertebral segments of that region.

The Brachial Plexus.—The brachial plexus (Fig. 73, No. 19) arises principally from the roots of the last three cervical and the first dorsal spinal nerves. These branches anastomose beneath the deep face of the scapulo-humeral articulation.

The brachial plexus divides into two distinct parts. The first (Fig. 73, No. 20) the dorsal called the superior thoracic is distributed to the serratus and the rhomboideus muscle. The second is the inferior thoracic (Fig. 73, No. 21), which consists of the main plexus, and which gives off muscular branches especially to the sterno-coracoideus and to the main portion passing on as the nerve trunk to the structures of the wing.

From the superior thoracic plexus there are several secondary plexuses formed.

The *dorsalis* or *serratus plexus*. This plexus is located adjacent to the anterior part of the main plexus and is formed of from two to four spinal nerve branches. The rhomboideus is supplied by branches which may be traced to the first root of this plexus. This nerve is called the rhomboideus superficialis. Another nerve is given off from the middle part of the dorsal side of the plexus and is called the rhomboideus profundus. A third, the superficialis serratus, is the largest of the branches. It breaks up into terminal branches, one going to each serration, or digitation, of the serratus muscle. Branches from the anterior portion of this plexus are distributed to the patagii muscles.

The superior brachial plexus gives off the following nerves:

The *subcoraco-scapularis*, which is purely a motor nerve, springs from the anterior roots of the plexus. It gives off a branch to the following three muscles: subcoracoideus, subcoracoido-scapularis, and scapulo-humeralis.

The *scapulo-humeralis* is distributed to the scapular humeral region.

The *latissimus dorsi* is located on the peripheral border of the scapulo-humeralis or supraspinatus. It originates from the second, and the fourth nerve roots of the plexus. It is located on the dorsal side of the plexus. It divides into two main branches, one going to the latissimus dorsi and the other to the scapulo-humeralis muscle. From the side are given off branches which supply the patagii muscle and enter into the formation of the dorso-cutaneous plexus.

The *axillaris* springs from the second and the third root of the brachial plexus, extends in a lateral direction, passing the ventral and distal rim of the insertional part of the posterior scapulo-humeralis or teres et infraspinatus and enters near the capsular ligament of the shoulder-joint. It gives off to this joint a small branch, called the recurrent articularis, which passes outward between the triceps brachialis and the humerus. It lies on the inner side of the major deltoid, and patagii muscles, and also the skin of the lateral shoulder and the upper arm region. The recurrent axillaris communicates with the main branch of the radial nerve.

The *cutaneus brachii superior* is a small nerve which springs from the last brachial plexus root and passes between the skin and the triceps brachii muscle on the dorsal surface of the upper arm, or brachial, region. It extends down over the elbow region, where it gives off numerous branches to the skin of these regions, to the meta-patagium and to the extensor muscles of the upper arm.

The *brachialis longus superior* is a large nerve trunk which springs from most of the other plexus roots. It extends around to the dorsal side of the upper arm and supplies the skin, the feathers, and the muscles of the forearm and the hand. Branches from this nerve trunk supply the triceps brachii and other muscles of the region. The main branch passes between the radius and the ulna, and, passing the elbow-joint, gives off branches to that joint. Passing on, it divides into two branches. One of these extends superficially over the upper part of the condylus ulnaris and supplies the extensor digitorum communis and the extensor metacarpi ulnaris, or flexor metacarpi radialis muscles, and, continuing superficially to the ulnar side, is distributed to the skin of the region. The second, a deep branch, extends on the ulnar side of the radius over the extensor indicis longus, and innervates the latter muscle, the extensor pollicis longus, the extensor pollicis brevis, the adductor pollicis, the interosseous palmaris, and the flexor digitorum.

The inferior brachial plexus gives off the following branches:

The *supra-coracoideus*, a large nerve which springs from the first main root of the plexus, and passes outward through the foramen coracoideum of the sternal ligament.

The *sterno-coracoideus*, a small nerve which extends downward from the plexus.

The somewhat large *posterior coraco-brachialis*, nerve which springs from one or two middle roots of the plexus and accompanies the pectoral nerve.

The *anterior thoracic*, a large nerve which springs from two or three of the posterior roots of the plexus and extends to the shoulder cavity where it branches. The anterior branch is distributed to the patagium and the front part of the pectoral muscles. The posterior branch supplies muscles along the side of the thorax and extends into the abdominal muscles.

The *anterior coraco-brachialis*, a small nerve which springs mostly from the inferior longus brachialis, passes to the distal end of the tuberculum humeralis radii, then passes backward between the front part of the humerus and the biceps, where it supplies the anterior coraco-brachialis.

The *cutaneous brachialis* et *inferior brachialis*, a small nerve which springs from the posterior roots of the plexus. It is distributed to the skin of the region and branches are given off to the patagii and the ventral wing surface. A few branches extend as far as the upper arm.

The *brachialis longus inferior* (Fig. 68, No. 2), a continuation of the main trunk of the inferior brachial nerve, which comes out of all the plexus roots except the first. It gives off some branches to the pectoral region and then enters the shoulder cavity. It gives off the anterior coraco-brachialis, passes down the humerus in an S-shape, and divides into two branches, the ulnar and the median.

The *ulnar nerve* (Fig. 67, No. 3) divides into numerous branches which are distributed to the ulnar side of the forearm and the hand. The ulnar nerve passes below the skin of the ulnar outside rim of the forearm and gives branches to the carpi ulnaris, finally supplying the flexor digitorum; it then passes with the tendon of this muscle downward to the interosseous dorsalis, the abductor indicis, or flexor minimi brevis, the flexor pollicis, and the abductor pollicis, or extensor proprius pollicis.

The *median nerve* (Fig. 67, No. 2) extends down the arm and gives branches to the biceps muscle, and to the patagii region. Continuing, the median nerve supplies the pronator muscle and the brachialis inferior, or brachialis anticus, and then divides into two branches. The first and largest branch passes between the two pronator muscles giving branches to the pronator profundus, or pronator longus, and the flexor digitorum profundus muscle.

This nerve then passes anteriorly along the tendon to the base of the digit, and, on the dorsal side, joins the other branch of the median nerve. At this point there is often a plexus, and at the point of fusion of the ulnar nerve on the ulnar side there may also be a small plexus. These plexuses mainly supply the skin of these regions. The second branch, passes forward, crosses the extensor carpi radialis and lies just below the pronator profundus, or pronator longus, giving branches to the skin. It passes downward on the radial side of the ulnar and supplies the muscles of that part; it then continues down the hand. Branches are given off to the extensor proprius pollicis, the interosseous dorsalis, the flexor digitorum, and to the skin between the thumb and the forefinger.

The *intercostal nerves* (Fig. 70, No. 18) are given off from the spinal cord of the dorsal region. The superior branches are small and supply the superior dorsal region. The inferior branches lie one behind each rib, innervate the intercostal muscles, and give a few twigs to the superficial thoracic muscles.

The Lumbo-sacral, or Crural, Nerve Plexus.—The crural plexus (Fig. 70, No. 35 and 36) is made up of trunks from the last two lumbar and first four sacral spinal nerves. There are two portions of this plexus separated by a considerable distance.

The anterior portion (Fig. 70, No. 35) consists of the lumbar nerves and a portion of the first sacral nerves. The fusion takes place on the bony ridge that separates the lumbar from the sacral region.

The posterior portion (Fig. 70, No. 36) consists of a part of the first sacral and all of the three succeeding nerves.

The anterior nerve of this plexus is distributed to the abdominal muscles. An anterior branch is given off to the sartorius. A large cutaneous branch enters between the sartorius and the ilio-trochanteric eminence, and supplies the outer and the upper surface of the upper thigh region. Several short branches are given off from the middle, or main, mass of the crural plexus to the ilio-trochanteric, or gluteal, muscle. Another nerve passes over the side of the sartorius adjacent to the femoro-tibialis, or extensor femoris, and supplies the ilio-tibialis, or gluteus primus. The rest of this section of the crural nerves pass in a distal direction over the inner and front side of the cutaneous nerve and enter the ilio-femoralis, or gluteus medius, the ambiens and the femoro-tibialis muscle.

The *furcalis nerve* comes out between the last two lumbo-sacral vertebræ.

The *obturator nerve* springs from several roots. The anterior root of this nerve comes from the main trunk of the crural plexus, and its last root from the furcalis nerve. The obturator nerve extends in a ventral direction from the plexus, then horizontally on the inner surface of the abdominal cavity. It gives off twigs to the obturator muscle and then passes through the obturator foramen. After leaving the abdominal cavity it gives branches to the accessorius, the obturator, and, the pubio-femoral, or adductor longus muscle.

The crural plexus gives off another large nerve trunk which extends downward. Shortly after emerging from the plexus it gives off a muscular branch to the ilio-femoralis, or gluteus medius muscle. It then passes between this muscle and the shaft of the femur, to the inner posterior surface of the thigh, and to the inner surface of the knee. It gives off a branch to the structures of the knee. At the knee this nerve terminates in several branches, some of which pass to the inner surface of the tibial head, the internal lateral ligament, the periosteum, the internal condyle, and finally to the upper part of the head of the gastrocnemius muscle.

The main part of this nerve passes downward as the cutaneous nerve, along the inner surface of the lower thigh.

The *ischiadic plexus* (Fig. 64, No. 41) has five to six roots, which fuse into the ischiadic trunk. This main trunk extends out of the abdominal cavity, through the ischiadic foramen, close behind the anterior trochanter of the ilium (Fig. 65, No. 1). It gives a branch to the external ilio-femoral or gluteus medius, and a branch to the post-acetabular portion of the ilio-tibialis, or gluteus primus muscle. The main trunk (Fig. 69, No. 13) passes downward to the lower portion of the thigh. In this course there are two branches which run parallel to the femoral artery and the femoral vein. It gives off a small muscular branch to the accessorius, and further down gives off a long slender branch to the outside of the knee-joint. It gives off a lateral cutaneous branch (Fig. 66, No. 7) to the posterior outer portion of the lower thigh. It supplies motor fibers to the external head of the gastrocnemius muscle.

At the knee region the ischiadica divides into three parts. The largest branch (Fig. 69, No. 14) passes with the tendon of the ilio-fibularis, or biceps femoris muscle, through a loop, the biceps band, and lies on the upper lateral surface of the fibula. It is covered by the external head of the gastrocnemius muscle. It innervates the three posterior muscles of the lower thigh, and then divides into two branches. One, the superficialis peroneus (Fig. 69, No. 16) passes with the profundus nerve, forming a double trunk and occupying the tibio-fibular groove on the antero-lateral side. It passes over the transverse ligament and the tibio-metatarsal joint, and after sending small branches to the structures of the tibial side of the metatarsus, it ends as a cutaneous nerve on the upper side of the third and the fourth toe.

The other branch is the *peroneus profundus*, which separates from the peroneus superficialis and passes downward in company with the tendons, under the transverse ligament, and then along the anterior upper surface of the metatarsus, where it innervates the muscles of that region. It gives branches to the malleolus, to the tendons of the third toe, and to the median part of the second toe, and supplies the cutaneous structures of the third and the fourth toe.

The third branch of the *ischiadica*, a long nerve, is given off just after the ischiadica passes through the loop. It passes downward between the two peroneal nerves. It is covered by a sheath. It passes over the posterior outside rim of the intertarsal joint and innervates the tendon sheath. The main portion of this nerve is located on the anterior surface of the tendon Achillis, and passes down on the plantar side, and innervates the periosteum and all plantar foot muscles. It finally radiates to the plantar surface of the three anterior toes.

The integument of the toes is sparingly supplied with nerves.

The nerve trunks that do not pass through the biceps band, or loop can be divided into a medial (Fig. 69, No. 15) and a lateral (Fig. 69, No. 17) portion. The medial portion (Fig. 69, No. 15) soon divides into numerous branches which supply the muscles of the posterior and the inner portion of the thigh. A rather large, long branch passes downward along the tendon of the plantar muscle, which lies on the posterior median edge of the tibia, and gives off twigs to the median and posterior part of the intertarsal joint, supplying the periosteum and other structures and the adjacent skin. It passes downward along the outer side of the medial metatarsal insertion of the tendon Achillis.

The *fifth branch* is given off from the second trunk and lies laterally. It is covered by the external head of the gastrocnemius, passes along the vena saphena, and gives off a short main branch to the inner side of the intertarsal joint. The main part passes the tibial side of the joint, becomes subcutaneous, and finally innervates the two plantar muscles. A lateral ascending inner branch innervates the flexor perforatus digitorum, and, in company with an outer branch, the external head of the gastrocnemius, and also the flexor pedis perforatus.

The *plexus pedundus* is formed from the spinal nerves coming out of the plexus ischiadicus. These fibers emerge caudalward and are directed horizontally. They frequently anastomose with each other, especially on the pubic rim and on the outside of the plexus ischiadicus. These branches are deeply imbedded in the kidney substance and innervate the pubio-coccygeus, or depressor coccygis lateralis, the ilio-coccygeus, the transversalis, the sphincter, and other muscles of this region, and the skin of the anal region.

THE BRAIN

The Brain Coverings.—The cerebro-spinal axis of birds is similar to that of mammals. The *meninges of the brain* are three in number, dura mater, arachnoid, and pia mater. The *dura mater* is the thickest. It is constructed of white fibrous connective tissue, and lines the cranial cavity. Thus it serves as an internal periosteum. It is continuous with the spinal dura mater at the foramen magnum, and is also prolonged as a sheath of the nerves. In birds of flight where the air-sacs and reservoirs are developed to the highest state, Sappy finds that, "just as the medullary tissue is replaced by air in the bones of birds, so might it be imagined that the subarachnoid fluid is also replaced by air around the spinal cord," and observations justify the correctness of this statement. The dura mater measures exactly the volume of the marrow in birds; so that there does not exist between the fibers and the nervous surface any space for an accumulation of liquid. This anatomical fact is sufficient to demonstrate the absence of subarachnoid fluid in the bird. In denying the existence of this fluid, it ought to be added that in this class of vertebrates, the spinal prolongation is covered by a triple envelope; and that between the pia mater and the dura mater is a thin transparent membrane, which is lubricated by a serous fluid. This fluid however does not collect; it only moistens the arachnoid membrane.

The *arachnoid* is located between the dura mater above and the pia mater below. The pia mater adheres closely to the nerve tissue.

The *falx cerebri* exists in fowls and in turkeys. It has the form of a segment of a circle, and extends from the middle of the interval of the openings for the olfactory nerves to the tentorium cerebelli. The falx cerebelli is absent. The tentorium is small and is sustained by a bony plate, and there are in addition two folds, one on each side, that separate the hemispheres from the tubercula quadrigemina. Owing to the absence of the falx cerebelli, the meninges lie close together. The falx cerebri is ossified in birds.

The Brain Structure.—The brain (Fig. 62, *C* and Fig. 75, *A* and *C*) is made up of three principal parts: the cerebrum, the cerebellum and the medulla oblongata. In a fowl of medium size the brain weighs about 150 grains.

The pons varolii is absent in birds. The crura cerebelli (Fig. 75, No. 9) are immediately connected with the corpora restiformia. The lower face of the isthmus is convex posteriorly; in front, the tubercula bigemina (Fig. 75, No. 4) are united to each other by a comparatively large transverse cord, formed by the optic nerves intercrossing in the median line. The superior face of the medulla oblongata is depressed above to constitute a fourth ventricle; in front of this ventricle are the tubercula bigemina, or *optic lobes*. These two voluminous tubercles are separated from each other above, where they embrace the cerebellum, and are salient on the sides of the lower face. They are hollow internally, and communicate with the aqueduct of Sylvius. The thalami optici are not well developed.

A large *transverse fissure* divides the cerebrum from the cerebellum (Fig. 75, No. 5). The optic chiasm (Fig. 75, No. 15) behind which lies the hypophysis (Fig. 62, No. C, 20) covers the region of the middle brain. The large transverse fissure is the dividing line between the hemispheres and the optic lobes. If the hypophysis be removed there will be observed a slit surrounded by gray matter which is called the tuber cinereum et infundibulum (Fig. 75, No. 18). On the pars commissuralis and the after brain there are visible the roots of the fifth, the ninth, the tenth, and the twelfth pairs of cranial nerves. Close beside the median furrow which extends to the front part of the long furrow of the spinal cord there is observed the third pair of cranial nerves, which have their origin in the middle brain. The fourth pair of cranial nerves extend from the roof of the middle brain on both sides between the middle brain and the optic lobes. These nerves finally emerge and become visible on the ventral surface.

The sixth nerve is visible near the middle furrow and almost in the middle of the pyramids, and near the roots of the fifth, the seventh, the ninth, the tenth, and the twelfth pair of cranial nerves.

THE DIVISIONS OF THE BRAIN

(Thalamus

(Pineal body

(Infundibulum

(Hypophysis

Forebrain   (Optic tract and chiasm

(Cerebral hemispheres

(Olfactory lobes

(Third ventricle

(Lateral ventricles

(Peduncles of the cerebrum

Midbrain   (Optic lobes

(Aqueduct

(Medulla oblongata

Hindbrain (Cerebellum

(Fourth ventricle

The *medulla oblongata* (Fig. 62, No. *C*, 1; Fig. 75, *A* and *C*) terminates anteriorly in the pars commissuralis and the pars peduncularis. As the spinal cord approaches the head there is a gradual swelling, or lateral thickening, which merges into the medulla oblongata. The superior and the inferior surface of the medulla oblongata (Fig. 75, No. *A*, 1 and *C*, 17) are flattened. There is a shallow furrow and a slight swelling at the point where the hypoglossal nerve emerges. The central canal of the spinal cord gradually comes closer to the upper surface and communicates with the fourth ventricle, at which point the posterior raphe is shortened and the sulcus longitudinalis posterior becomes shallow. At the point where the central canal terminates in the fourth ventricle there is a V-shaped point called the *calamus scriptorius* (Fig. 75, No. 2). The *fourth ventricle* is located on the upper wall of the medulla oblongata below the cerebellum, and is bounded laterally by the peduncles of the cerebellum. It is marked posteriorly by the calamus scriptorius and anteriorly by the valve of *Vieussens*.

The *valve* of *Vieussens* is located at the posterior end of the aqueduct of Sylvius. The posterior part of the fourth ventricle is marked by grooves or furrows. Extending along the central part of the floor of the fourth ventricle there is a sulcus, or groove, called the sulcus centralis, which divides the superior pyramids. On the median lateral sides there are two points of gray substance which form the alæ cinereæ. The roots of the pneumogastric, the glosso-pharyngeal, and the spinal accessory may be traced to the alæ cineræ and the gray matter of the ridges of the medulla oblongata. The sixth pair of the cranial nerves emerge from the medulla near the sulcus centralis and to the side of the auditory nerve. Part of the trigeminus emerges from the rim of the furrow next to the ridge. At the outer border of the medulla are observed the thick ends of the roots of the pneumogastric, the spinal accessory, the glosso-pharyngeal, and the auditory facialis.

On both sides of the sulcus longitudinalis inferior are found the *inferior pyramids*, or *pyramidal columns*. These pyramids become expanded near the origin of the third pair of nerves and merge into the cerebral peduncles, or crura cerebri.

The pyramidal fibers may also be traced to the optic lobes. The roots of the abducens are found at a point between the crura cerebri and the optic lobes. The third pair lies to the side of this, and the roots of the trigeminus are adjacent to those of the third. A bundle of fibers from this region pass into the cerebellum and form the crura cerebelli; others pass into the cerebellum from the side, spreading out in fan-shaped radiation, and forming the white central substances peculiarly arranged, called the arbor vitæ, or *tree of life*. A third bundle fuse with the crura cerebelli anteriorly, and pass into the peduncles of the cerebrum, or crura cerebri.

The gray ganglionic substance forms columns which extend the whole length of the medulla oblongata and into the cerebri. The medulla oblongata contains numerous centers which preside over various visceral functions as deglutition, respiration, thermotactic, secretory, cardiac, and digestion.

In the medulla as in the spinal cord there are five main groups of cells. First, the posterior or upper horns, from which come the somatic sensory nerves. Second, Clark's cells, located centrally, which are the origin of the ganglionated splanchnic nerves. Third, cell groups of the lateral horns which are the center for the non-ganglionated splanchnic nerves, and certain other nerves for the viscera, including the enteric muscles. Fourth, cell groups of the anterior horns, which are the center of all somatic voluntary muscles. Fifth, groups of single cells probably belonging to the posterior horns which are probably centers for other splanchnic nerve fibers.

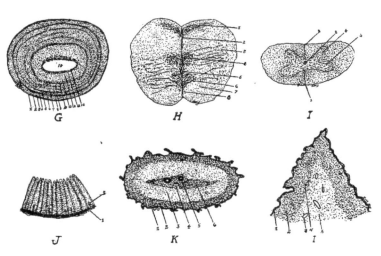

FIG. 77, *A*.

G, A section through the tuberculum bigeminum. 1, Pia mater. 2, Nerve fibers. 3, Fine granular ground substance. 4, Thin layer of small cells. 5, Fine granular ground substance. 6, A second thin cell layer. 7, Fine granular ground substance. 8, Third thin cell layer. 9, Fine granular ground substance. 10, Fourth widely extended cell layer. 11, A fine granular layer.

12, A thick layer of spindle cells. 13, A layer of medullated nerve fibers. 14, Ependemal cells. 15, The sinus.

H, A transverse section through the medulla oblongata, 1, 3, 4 and 6 are ganglionic centers. 2, The dorsal groove. 8, The ventral fissure. 5, The central canal. 7, Fiber tracts.

I, A transverse section through the cervical portion of the spinal cord. 1, Ventral septum. 2, Dorsal septum. 3, The central canal. 4, The dorsal horn. 5, The ventral horn.

J, A section through the wall of the oil gland, or rump gland. 1, The stroma. 2, The tubular glands some of which are branched.

K, A section through a spike of a comb of a cock. 1, The epithelial layer. 2, A dense fibrous subepithelial layer. 3, A second fibrous vascular layer. 6, A denser central core supporting the large arteries and veins. 4, A vein. 5, An artery.

L, A section through a lobe of the wattle of a hen. 1, The epithelial layer. 2, A dense vascular layer. 3, A vein. 4, An artery. 5, A loose fibrous vascular structure.

The *gray matter in the cord* is in the form of two commas placed with their backs together with the central canal passing between them. The central canal is located in the center of the cord. As the cord nears the medulla oblongata the form of the gray matter changes. The second, the third, and the fifth ganglion centers are arranged laterally, and show a distinct side-horn group. The central canal separates the two Clark's columns; later the side-horn group enlarges and extends ventrally around the anterior horn group and joins on the other side in half-moon shaped formation.

The *central ganglionic mass of the spinal cord*, the center for the enteric visceral system, is divided into three complex parts. First, there are cell groups on both sides of the posterior raphe. Second, the half-moon shaped ventral groups extending parallel peripherally described above. Third, lying between groups one and two a complex group, which possesses a large collection of cells. These furnish splanchnic nerve centers, the anterior supplying motor nerves and carrying motor impulses outward, and the superior sensory which carry impulses to the centers. The cell group of the superior horns takes a sidewise peripheral position (Gadow).

The *cerebellum* (Fig. 62, *C*, 2) is located above the medulla oblongata and posterior to the cerebrum. The cerebellum consists principally of a median lobe called the worm. The lateral lobes are conical and rudimentary. The under part of the worm forms the roof of the fourth ventricle.

On the upper surface of the cerebellum are numerous transverse markings (in the hen 13 or 14) which divide the lobe into leaves. When the cerebellum is cut lengthwise there is observed on the sectioned surface the peculiar arrangement of the white and the gray matter, the arbor vitæ, mentioned above. There is a small cavity in the cerebellum which communicates with the fourth ventricle.

The *cerebellar cortex* shows four parts, namely: First, the central part composed of white medullated fibers, between which are arranged neuroglear cells. Second, a rather reddish layer of cells of different sizes. These cells are embedded in a fine ground substance and are about 0.003 mm. in diameter. Third, a plain layer of large pear-shaped ganglion cells, the cells of Purkinje. Fine branches extend outward from the apex of these cells. Fourth, an outer gray layer which consists of small multipolar nerve cells with some neuroglear cells. The color of these three layers, according to Gadow, depends upon the color of the plumage; that is, if the plumage is dark, this layer is dark; if light, the layers are light. The marked coloring is said to be most distinct in the second layer.

The anterior portion of the cerebellum consists of medullated nerves whose fibers run crosswise. These are the extensions of the true cerebellar fibers, which become fewer and fewer as they proceed anteriorly. Finally, a few remaining fibers running crosswise fuse as the commissura Sylvii, and extend into the lobi optici. A layer of nerve cells extends under these medullated nerve fibers.

A bundle of fibers extending from the medulla oblongata. These fibers extend into the optic lobes, form the crura cerebri, or peduncles of the cerebri, and also form the inferior wall, or floor, of the aqueduct of Sylvius. The crura are sometimes spoken of as the partes pedunculares. These contain gray matter. The following division is made of the ganglion of this region. First, a group of cells near the base of the peduncles, which are divided from the ventral rim by the arciform fibers. This ganglionic formation may be considered an extension from the medulla oblongata. Second, a group of the ganglion cells of the lobus optici. Third, a group of cells lying near the lobus opticus, which give rise to the ascending roots of the trigeminus. Fourth, a group of ganglionic cells from which spring the roots of the motores occuli, and which lies near the middle line and under the sulcus centralis. Many ganglion cells are found to the right and to the left of the sulcus centralis, which show different arrangement of the cells at different levels. Fifth, a group of irregular cells which extends centrally on the superior walls of the third ventricle and to the superior part of the lobus opticus.

The microscopic structure of the roof of the lobus opticus, or corpus bigeminum, shows that the layers are arranged parallel to the pia mater and are divided into the following parts:

Externally, the pia mater.

An outer layer of very fine nerve fibers, which lies just below the pia mater.

A layer of fine granular ground substance.

A thin layer of small cells the diameter of each of which is 0.038 mm.

A layer of fine granular ground substance.

A second thin cell layer.

A zone of fine granular ground substance.

A third thin cell layer.

A fine granular ground substance.

A fourth widely extended cell layer.

A fine granular layer.

A somewhat thick layer of spindle-shaped cells.

A layer of medullated nerve fibers.

This last is the inner layer of nerve fiber. On the inner surface of this layer we find the ependymal cells, the cells that line the cavity of the optic lobes. The commissura Sylvii form the covering of the lobi optici and join these two lobes. This commissure is formed from the upper and the lower layer of nerve fibers. In the commissure these fibers cross each other.

The *hypophysis* (Fig. 62, No. *C*, 20) lies back of the chiasm, or optic commissure, and below the middle of the third ventricle. The infundibulum, a pedicle-like structure, connects the hypophysis with the third ventricle of the cerebri. It contains a cavity and forms an extension of the third ventricle. The hypophysis is divided into two lobes, an anterior and a posterior.

The *third ventricle* communicates anteriorly with the lateral ventricles of the cerebri through the foramen of Monro, and posteriorly with the fourth ventricle through the aqueduct of Sylvius. The posterior walls of the third ventricle are relatively thick and form the posterior brain commissure. The wall is thin from the commissure to the chiasm. Within it lie two ridges, which connect the two hemispheres with the corpus callosum, and, farther anteriorly, with the anterior cerebral commissure. The space from the third ventricle and to the anterior commissure is called the laminæ terminales.

The *pineal gland*, located between the cerebri and the cerebellum, lies against the choroid plexus, which covers the highest point of the third ventricle.

The *cerebrum* is divided by a deep longitudinal fissure into two hemispheres. The cerebrum is shaped somewhat like the heart on a playing card. The lower face is somewhat flattened. The upper and lateral sides are not provided with convolutions or sulci, but are smooth. The fissure of Sylvius is faintly marked on the inferior face. The olfactory lobes appear well developed and are relatively large for the size of the brain. They lie close together on the extreme anterior portion in the median line.

The *olfactory cerebral crura* emerge from the cranium at the upper angle between the posterior wall, the roof, and the septum of the orbit, and pass through the olfactory foramina and in grooves on the upper part of the septum; then passing forward they penetrate the frontal structure and are finally distributed over the turbinated mucous membrane.

The *corpus striatum* is large; it occupies nearly the entire floor of the ventricles.

The *anterior commissure* is found between the hemispheres. Its middle part lies in the lamina terminalis on the division between the anterior and the middle brain, and close to the thalami optici. Its side extensions are rounded masses called the nervi amygdales.

The *corpus callosum*, quite rudimentary, lies on the posterior dorsal rim of the anterior commissure.

The *corpus striatum* forms a thin broad ridge, which passes ventrally. This structure consists of twenty-five large pyramidal cells which lie in the posterior lateral dorsal section, and of from ten to fifteen pyramidal cells which are located in the rest of the section. In addition to these two groups there are many cells about 5 millimeters in diameter. These lie, six or more, in a nest imbedded in neuroglia. There is a thin layer of spindle cells near the ventral line. The outer nerve nest, or nerves amygdalis, lies in the posterior lateral ventral part. It is covered by a thin membranous layer. Its cells are pyramidal in shape and are from 10 to 15 millimeters in diameter. The cells terminate in spindle form toward the periphery. Nerve fibers extend from the anterior commissure.

The outer wall of the corpus striatum consists of the following layers:

An outer white layer consisting of fine medullated nerve fibers imbedded in a ground substance containing numerous nerve cells.

A layer of ganglionic cells, consisting of pyramidal cells from 10 to 15 millimeters in diameter, of other round cells 20 millimeters in diameter, and finally of cells only 5 millimeters in diameter. This layer forms a reddish line in the dorsal portion of the corpus striatum.

These layers form part of the median and the posterior cerebral wall of the lateral ventricle. The lateral ventricle is closed except for a slit-like opening behind the posterior commissure. This is the foramen of Monro, through which the lateral ventricle communicates with the middle, or third, ventricle.

The *choroid plexus* is found at the base of the lateral ventricle.

The wall of the lateral ventricle becomes thinner near the region of the transverse commissure on the surface toward the middle brain. At this point the pia mater and the ependyma, or the lining cells of the ventricular membrane, form the major part of the wall.

The domestic fowl does not have the hippocampus.

The wall of the middle ventricle consists of an outer white layer, which is arranged similarly to that of the corpus striatum. The ventral rim of the wall is formed by a spiral band which consists of fine medullated fibers, in which are imbedded a few cells.

The *processus cerebri mammillares* are also called the tubercula olfactoria. From these structures extend forward the olfactory nerves, the posterior roots of which may be traced to the walls of the lateral ventricles.

The structure of the processus cerebri mammillaris is made up of five layers as follows:

An outer layer, which consists of colorless olfactory fibers extending in all directions.

A granular ground substance layer, in which are imbedded a few cells.

A thicker granular layer, on the inner edge of which are twenty large pyramidal cells with processes pointing outward.

A layer of closely packed cells, in which are fine medullated nerve fibers. These cells are about 5 millimeters in diameter.

An innermost layer of ependymal cells.

Near the hemisphere the peripheral nerve fibers and nerve cells disappear, leaving only the ground substance of the processus to aid in the formation of the hemisphere.

There is a long bundle of fibers on the lower surface of the hemisphere, which blends with the substance of the olfactory fibers.

The peduncles of the cerebrum (Fig. 62, No. 21) are the slightly diverging columns of nerve tissue, which form the anterior continuation of the medulla oblongata and disappear under the optic tracts and the chiasm.

THE SYMPATHETIC NERVOUS SYSTEM

The sympathetic nervous system commences anteriorly at the large superior cervical nerve ganglion. This ganglion, anteriorly, brings into communication the glosso-pharyngeal, the pneumogastric, or vagus, and the sympathetic system.

The following nerve trunks emerge from the superior cervical nerve ganglion: a large trunk fusing with the vagus and directed downward, accompanying and surrounding the carotid artery; a second large trunk fusing with the recurrent pharyngeus and the glosso-pharyngeus; a third trunk which merges with the hypoglossal; finally several trunks which are distributed to the head.

The *temporo-lacrimalis*, one of the large sympathetic nerves of the head, as it extends from the cervical nerve ganglion, receives branches from the ganglion radicis vagi. This nerve, passing between these two ganglions, extends horizontally forward and outward through a foramen, crossing the glosso-pharyngeus. Near the Fallopian canals it crosses the facial nerves, lies supero-laterally, and receives a short branch from the facial nerve whose fibers are traceable to the geniculate ganglion. It also receives a branch from the recurrent maxillaris.

This nerve accompanies the external ophthalmic artery forming around it a network of fibers, called the external ophthalmic plexus. It lies outside and downward from the optic nerve, sends fine branches to the external ophthalmic artery and to the masseter artery, and extends along with a small branch of the superior maxillary nerve to the skin of the outside rim of the eye cavity.

The *ophthalmic plexus*, a second trunk extending to the head, enters in its course, anteriorly, the lacrimal plexus. Its fibers also supply the lacrimal gland, and finally anastomose with the second recurrent branch of the trigeminus.

The sympathetic *caroticus cephalicus* nerve, a third trunk extending to the head, after emerging from the large superior cervical nerve ganglion, receives some small branches from the glosso-pharyngeal ganglion, and then enters, in a horizontal manner, a foramen, the canalis caroticus externus, located in the lower part of the basi-occipital and the sphenoid bone. The anterior opening of this foramen, or canal, is close to the posterior part of the pterygoid bone. Inside this canal the sympathetic caroticus cephalicus receives a small branch which extends in a straight line from the basal part of the facial nerve. After receiving this branch, the nerve trunk, passing to the ear drum, lies close to the petrosum and the sphenoid. It is covered by the masseter muscle. The carotic ganglion is located at the point of the fusion of the caroticus cephalicus with the main trunk. At this point the caroticus cephalicus divides into two branches. The first is the superior recurrent nerve, which lies close to the upper surface of the alæ of the sphenoid, passes between the obliquus externus and the orbital wall, around the eyeball, and finally into the orbital septum and the internal muscles of the eye. It communicates with the ophthalmic nerve and sends twigs to the lacrimal gland, the gland of Harder, the upper eyelid, and the nose. The orbito-nasale ganglion is located where these nerves communicate near the nasal region. The second branch of the sympathetic caroticus cephalicus is the inferior recurrent nerve. This branch passes forward and dorsally from the pterygoid bone to a point where the upper rim of the rising wings of the jaw bone meets the sphenoid rostrum. In this course, some branches are given off to the pharynx, one branch near the lacrimal gland, and a branch which communicates with the superior maxillary nerve just before the nerve enters the jaw bone. The spheno-palatine ganglion is at the point of fusion.

The terminal branches of the spheno-palatine are distributed to the hard palate, the nose, and the lacrimal gland.

From the large cervical nerve ganglion and near the roots of the caroticus cephalicus are given off a few small nerve fibers which pass alone to the pharynx or accompany the jugular nerve branches, and fuse with the main trunk of the caroticus cephalicus nerve.

From the large superior cervical nerve ganglion the sympathetic trunk extends downward toward the thorax. It is covered deeply with muscles. This portion is known as the cervical sympathetic nerve trunk. A trunk lies on each side of the cervical vertebræ. The large thoracic nerve ganglion, the inferior cervical, is located along this trunk at the entrance of the thorax. From this ganglion there is given off the recurrent cardiacus which supplies the heart. The end branches of the sympathetic nerve trunk blend with the pneumogastric nerve. The inferior nerve ganglion is the ganglion cardiacum. Near this ganglion is found a nerve plexus in which there are imbedded peripherlistic ganglia. This thoracic plexus, accompanying the collica artery, passes to the abdominal region, supplying the intestines and taking part in the formation of the abdominal plexus. The abdominal plexus is located near the anterior portion of the kidneys. Its fibers are directed mainly downward to the visceral organs. The large intestine, the rectum and the copulatory organs receive branches. These latter branches take part in the formation of the pedunda nerve plexus. Some branches of this plexus, follow the branches of the posterior mesenteric artery.

The *thoracic trunk* (Fig. 64, No. 42) of the sympathetic is double. The anterior portion gives off an anterior splanchnic nerve, or plexus (Fig. 64, No. 43) which accompanies the celiac axial artery to the gizzard and liver, communicating with the pneumogastric. The posterior splanchnic nerve is intimately combined with the adrenal body, and the testes, or the ovary (Fig. 64, No. 45). Intestinal branches accompany those of the mesenteric arteries (Fig. 64, No. 46). Other branches supply the kidneys, and communicate with long branches of the spinal nerves destined for the cloaca and adjacent parts, and thus form a plexus similar to that found in mammals.

FUNCTIONS OF THE NERVOUS SYSTEM

According to function, the nerves composing the trunks are divided into afferent and efferent.

The *afferent nerves* are those that convey the impulses from the periphery of the body to the nerve center, which are located in the brain or in the spinal cord. The impulses conveyed are those of special senses, as sight, hearing, taste, touch, and smell. Impulses producing sensation pleasurable or painful come from the skin, the muscle, or the viscera.

The *efferent nerves* are those which convey impulses from the nerve centers to the periphery. These impulses may be motor as those going to the muscle cells of the skeletal muscles, the viscera, or the blood-vessels. These motor impulses make movements in these organs possible. In the blood-vessel they result in the control of the caliber of the vessel. These impulses may be of an inhibitory character, as in slowing the heart. They may be secretory impulses stimulating the gland to activity or regulating metabolism.

The *ganglia* are nerve centers which receive and generate impulses; the nerve trunks are filaments which convey impulses. The gray matter of the cord described above is the ganglionic portion, and the outer white matter is made up of nerve fibers which convey impulses from one part of the cord to another, or to and from the brain.

The nerves that have their roots in the spinal cord superiorly, are sensory; that is, they convey the sensory impulses from the periphery to the cord ganglion and to the brain. They have a ganglion just outside the cord.

The nerves that have their roots in the spinal cord inferiorly; that is, they convey motor impulses from the nerve centers to the periphery. The function of the inferior roots is to supply all the voluntary muscles as well as the oviduct, the intestines, and other hollow viscera, including the blood-vessels with the power of movement. Many of these fibers pass to the sympathetic ganglion and are distributed as sympathetic nerve fibers.

The cord is divided into different tracts, that is, certain groups of fibers convey certain kinds of impulses.

The superior column of the cord conveys to the cerebrum such impressions as temperature, pressure, and muscular tension.

The fibers of the lateral columns carry sensations of pain.

The fibers of the direct cerebellar tract carry impulses which result in the maintenance of the equilibrium of the body.

All the voluntary impulses originate in the cerebrum, pass through the cerebellum and travel direct to the bulb; they then pass over to the opposite side, and travel by the crossed pyramidal tract to the multipolar cells of the inferior horn of the spinal cord, and transmit the impulse through motor fibers that originate at that point in the cord.

All sensory impulses enter the brain on the side opposite their origin, and all motor impulses leave the brain on the side opposite that to which they are distributed. Injury of the motor area of the right side of the brain leads to paralysis of the left side of the body.

An impulse of the vasomotor nerve travels in the lateral column of the cord.

A nerve impulse may originate in the brain and be modified in passing through a ganglion in the spinal cord or in the sympathetic system.

The system of *reflex action* is as follows: first, an efferent nerve which conveys the impulse from the periphery to a nerve center; second, a ganglion, or nerve center, to receive the impulse and generate other impulses; third, an efferent nerve to convey the impulse from the nerve center to the periphery.

The following is an example of reflex action. The foot of a fowl is pierced with a pin. The sensory impulse is conveyed by sensory nerve fibers to the nerve centers. In this center the ganglionic nerve cells generate a motor impulse which is sent back through the motor nerve fiber to the muscles controlling the part. The result is that the muscle contracts and jerks the foot.

There are many *functional centers* located in the medulla oblongata. Destruction of this part of the system results in instant death of the bird. In addition to furnishing a path for fibers carrying impulses from the body to the cerebrum, it furnishes a large number of centers for such functions as respiration, swallowing, secretion, temperature, and vasomotor and cardiac activity.

The *function of the cerebellum* is principally that of coördination. It brings about harmony and rhythm in muscular movements. If the cerebellum be removed, the bird can no longer walk. It has lost, with this removal, all power to coördinate.

The *cerebrum* of the bird has no convolutions, and the gray, or ganglionic, portion is thin, indicating low power of intelligence. There are certain areas presiding over other functions, as motor areas, sensory areas, and so on.

A careful study of the brain of the fowl shows us that the centers presiding over sight and smell are well developed.

The olfactory bulbs are the centers of the sense of smell.

The optic thalamus is the center of the sense of sight.

The sympathetic nerve system transmits impulses to the involuntary muscular structure of all organs, including those of the intestinal tract, the blood-vessels, and perhaps also the glands.

ESTHESIOLOGY

THE SENSE ORGANS

The five special senses are seeing, smelling, tasting, hearing, and feeling.

THE ORGAN OF SIGHT

The sense of sight in the bird is well developed. The eye (Fig. 26, *D*, *E* and *F*), the organ of sight, is relatively large, round laterally, and rather flattened antero-posteriorly. It is located at the side of the head. The eyeball is only slightly movable. The septum interorbital separates the two eyeballs laterally. As in mammals, the eyeball has three coats, which are from inside to outside, the retina, the choroid, and the sclera. The **sclerotic coat** is completed anteriorly by the cornea with which it forms a union called the *corneo-scleral juncture*. Around the cornea the sclerotic coat contains a ring of osseous scales varying in number from twelve to twenty. The sclera may become ossified posteriorly, forming an osseous sheath around the optic nerve. The *pecten* is a vascular comb-like membrane stretching from the nervous opticus to the crystalline lens. The choroid coat is always black. The pupil in the hen is also black and round. The iris contains striated muscular fibers. The *membrana nictitans*, located at the inner angle is well developed. It is moved by two muscles (Fig. 26, No. *B*, 7, 8, 10). The lacrimal gland and the gland of Harder are present. There is no meibomian gland.

The *tears* are secreted by the lacrimal gland and are drained away from the fore part of the eyeball by two small canals which extend into a lacrimal sac. From this sac there extends a tube into the nasal cavity called the lacrimal duct.

The lower lid is the larger and often incloses a small cartilaginous plate. The *conjunctiva* is a true mucous membrane which covers the anterior portion of the eye cavity attaching to the cornea-scleral juncture.

The **choroidea** is rich in pigment. On its inner surface lies the dark pigment layer of the retina. The corpus ciliare, that part of the choroid coat bearing the ciliary processes, consist of numerous folds. The ciliary muscles are arranged obliquely. Each consists of three digitations.

The *iris* is covered on the posterior side with pigment, the color of which determines the color of the eye. The yellow pigment of the iris has been said to be due to carotin and xanthophyll and the black pigment to melanin. The enlarging and the contracting of the pupil is brought about by muscles. The reduction of the pupillary caliber is due to the sphincter pupillary muscle. It is said by some anatomists that the muscles controlling the caliber of the pupil in the bird furnish voluntary motion and that the capability of accommodation of the eyes is greater in birds than in mammals.

The **retina** contains no blood-vessels; otherwise the structure is similar to that of mammals. The crystalline lens is flattened on the corneal side and is convex posteriorly. The lens epithelium develops into fibers in the parts close to the equator, and are almost perpendicular to the eye axis. The corneal portion is relatively small.

The **sclerotic coat** is dense and white. It is divided into three layers. It is thin and flexible, and somewhat elastic posteriorly. It has an internal layer of hyaline cartilage. Anteriorly its form is maintained by the circle of osseous plates mentioned above. These plates, interposed between the exterior and the middle layer, are located immediately behind the cornea. The scales are thin and of oblong quadrate shape, being elongated from before backward.

The choroid coat is a membrane loosely cellular and highly vascular. It is impregnated by a black pigment. Opposite the bony circle the choroid separates into two layers. The external layer is the thinner and adheres at first firmly to the sclerotic; it passes forward to become continuous with the iris. The inner layer is thicker than the external. The two layers are made up of radiating fibers which terminate anteriorly in the ciliary processes, the ends of which are adherent to the capsule of the crystalline lens.

The iris is delicate in structure. It is composed of a fine network of interlacing fibers.

The ciliary nerves and blood-vessels run in the form of single trunks between the choroid and the sclerotic, and terminate anteriorly in a ring-shaped plexus for the supply of the iris and the muscular circle of the cornea. As stated before, the pupil in the fowl is round, but in the goose it is elongate transversely; and in the owl, a vertical oval.

The optic nerve approaching the sclerotic coat becomes altered into a conical extremity, which enters a sheath and is directed downward and obliquely forward. The extremity of the optic nerve in the interior of the eye presents a white narrow streak. Branches of the opthalmic artery enter the eye between the lamina of the retina, along the whole extent of the oblique slit, and penetrate the fold of the pecten upon which they form a delicate ramification.

The crystalline lens is of soft texture. It is inclosed in a capsule and is nearly round. It adheres very firmly in the depression in the anterior part of the vitreous humor. The capsule is lodged between two layers of hyaloidea, which as they recede from each other, leave around its circumference the sacculated canal of Petit.

The cornea is of horny consistency and is transparent. Light thus rapidly passes through it to the posterior part of the eye.

The vitreous chamber, lying back of the crystalline lens, contains a clear jelly-like substance.

THE ORGAN OF HEARING

The ear, the organ of hearing, has in the fowl no conchal cartilage. The *external auditory meatus*, or canal opening, is found on each side of the head, and is usually guarded by a few stiff, short feathers. In some kinds of birds these feathers are capable of being erected so as to direct the waves of sound into the inner ear. The outer entrance of the ear contains glands. This canal is short. It leads to the drum, which is somewhat convex from the outside, and which has a membrane forming a complete curtain stretched over the outer part. The irregularly formed drum has connection with the air-containing cavity of the skull, and by a thin cartilaginous canal, the Eustachian tube, with the pharynx. The *auditory ossicles* are represented by only a single bone, called the *columella*, which most closely represents the stapes of mammals. This is attached by processes of cartilage to the tympanic membrane. Owen considers these cartilages as representing the malleus of mammals. Huxley on the other hand considers them to represent the incus. Originating from the processes is a small muscle which is attached to the drum. This is by Shufeldt considered the Tensor tympani. The drum cavity through the fenestra vestibularis and cochlearis is connected with the labyrinth.

The **inner ear** consists of a membranous labyrinth, surrounded by a spongy bony structure—the bony labyrinth. In it are recognized the vestibulum, the three half-circle shaped canals, and the cochlea. The superior semicircular canal is the largest. The acoustic nerve enters at the end of the canals near the ampullæ. The nerves are supported in delicate vascular membranes lining the canals and slightly projecting into the ampullæ.

FIG. 78.—Diagram of the inner ear. 1, The integument. 2, The superior semicircular canal. 3, The external canal. 4, The horizontal canal. 5, The

ampulla. 6, The obtuse osseous conical cochlear cavity. 7, The Eustachian tube. 8, The tympanum. 9, Filaments of the auditory nerve.

The **vestibulum** is a small irregular cavity which communicates with the arcades and the cochlea, and through the fenestra vestibularis with the drum cavity. The endolymph of the vestibulum contains microscopic crystals of calcium carbonate. The semicircular canals are relatively larger and thicker than in mammals. The ampullæ in the upper and back part are separated by walls. The cochlea is an obtuse conical tube-like structure slightly curved at the blind extremity with the concavity directed backward. It contains a membranous lining. At this point the cochlea is broadened and accommodates a branch of the auditory nerve. This nerve spreads out in fine filaments upon the surface of the tubes. The hollow space of the cochlea is divided by a spiral partition, making two chambers, the scala vestibuli and the scala tympani. These walls extend from the beginning of the cochlea.

The **tympanic cavity** is formed by the occipital, the basi- and alisphenoids, the petrosal portion of the temporal bone. It represents the stapedial canal leading to the foramen ovalis and the pneumatic apertures by which the air from the Eustachian tube is conducted to the precranial diploë.

THE ORGAN OF SMELL

The nose is the seat of the peripheral portion of the organ of the sense of smell. The terminals of the olfactory nerves, which receive the impressions of odors, are broadened in the mucous membrane of the walls of the anterior nares. There are no ethmoidal volutes, or sieve-like structure, in birds. The nerve extends from the anterior portion of the brain in cone-shape, finally dividing into filaments which are distributed over the mucous surface of the turbinated bones.

THE ORGAN OF TASTE

The most important part of the organ of taste is the *tongue*. In birds the dorsum of the tongue is covered by a thick stratum corneum, a heavy layer of stratified squamous epithelium. The tongue, therefore is not in birds so well adapted for the perception of taste as it is in mammals. The lingual branch of the trigemini is lacking in birds. This fact makes the ninth pair, or the nervi glosso-pharyngei, the exclusive nerves of taste. There is an opinion ventured by one anatomist that, since the first and the second branch of the fifth pair, or trigemini, have terminal filaments in the hard palate, they may furnish fibers for the sense of taste. There are many taste cells on the tongue and on the dorsal palate.

THE ORGAN OF TOUCH

The peripheral parts of the organ of touch in the fowl are the skin and the feathers.

In a few birds special touch and taste perception can be supplied by the edge and point of the beak.

In the skin of birds are found numerous sensory nerve endings for tactile sense.

The sensory nerves, which provide the sense of touch, usually terminate in one of two forms. The first are the Herbst's bodies, which in many respects are similar to the Picinian bodies.

Herbst's *touch corpuscles* are found on all parts of the skin; they are especially numerous in the region of the tail and of the wing. In the wing they are particularly numerous in the region of the flight feathers. They occur in large numbers in the periosteum of the anterior tibial region and in the mucous membrane of the cloaca and of the generative organs. They are numerous in the conjunctiva and on the surface of the tongue. They are also found in the gums and in the beak.

Herbst's corpuscle is made up of a central fiber-shaped part with a smooth extension of the axis cylinder of its nerve (Fig. 29, G). This central fiber is surrounded by a peculiar protoplasmic body. Outside of this there is a double row of cubical cells which surround the axial part. These are close together. Outside of these there occurs a concentric lamellar layer, which contains cells. In the periphery these lamellæ become more distinct and contain larger but fewer cells. The capsule is made up of very thin layers, which are continued into the perineural layer. Each body contains a thin outer zone. The nerve fiber passes in a regular manner from the axis cylinder, the myelemma, and the sheath of Schwann, and as a delicate nerve twig is surrounded by a sheath consisting of several perineural layers. As it enters the center of the touch corpuscle, it loses its myeline sheath near the base of the body, and the terminal fiber becomes flattened. Its rim is directed toward the two rows of cubical cells, and it ends in a rounded knob.

The touch corpuscles are the largest in the mucous membrane of the cloaca and smallest in the skin. Hess has found these touch corpuscles in the large filiform papillæ on the side of the tongue and a few on its lower surface. They are also found in the soft skin of the edges and inner borders of the beak.

STRUCTURE OF APPENDAGES

The *skin of the fowl* is very thin and does not contain oil glands or sweat glands. In the fowl the oil is supplied by a tail or rump gland, the *glandula uropygii* (Fig. 35, No. 16). This gland is round or oval in shape, and about the size of a pea and consists of two lobes. It is of the tubular variety with a teat, which, in most instances has two openings. A medium septum divides the gland into two halves. The oil secreted by the columnar epithelial cells is collected in a body cavity located in the center of the gland; this has a duct extending to the surface. The bird by squeezing out a quantity of this oil into its beak oils the feathers, passing the beak over them, one by one. The oil renders the feathers practically impervious to water. It is necessary for the bird to give proper attention to its plumage in order that the feathers appear in prime condition. Should there be a disease of the oil gland, or should the bird become ill and neglect its toilet the result is an unkempt appearance of the plumage, the feathers becoming rather rough and more or less injured by the weather.

The *subcutis* is well developed and furnishes to the cutis great capability for movement, which is necessary for the rising and falling of the feathers. The corneum is very thin. Papillary bodies are present only in a very few places, such as the region about the eye and on the toes. Where the toes touch the ground in walking there are large wart-like thickenings of the epithelium. In most birds the shanks are unfeathered. The epidermis on the feathered parts of the skin is thin, dry on the surface, and abounds in continuous scales. The stratum corneum is very thick on the horny sheath of the beak, on the top of the toes, on the spurs of the cock, and on the scale plates that cover the skin on the shanks. The feathers of birds serve the same protective purpose as hair on mammals. In cold weather the skin muscles controlling the feather movements contract; thus the feathers become ruffled much as we observe the hair standing erect on horses under similar conditions. By increasing the dead air space around the body, the radiation of heat is retarded, and the body kept warmer. The corneum of the skin is not usually rich in blood-vessels; however, in the domestic fowls there is in the comb, in the wattles, and similar appendages of the head, a thick vascular network.

The *beak* and the *claws* are modified skin; they are true horn material. At the base of the beak there is often formed scales, which surround the nostrils, in whole or in part, and have a naked, or waxy, appearance, which is known as the *cera*.

FIG. 79.—Photomicrograph of the section of skin from the sole of the foot of a hen. 1, Horny stratified squamous epithelium. *a*, Stratum corneum. *b*, Stratum lucidum. *c*, Stratum germinativum. 2, Connective tissue supporting membrane, pars reticularis. 3, Blood-vessels.

The feathers may be considered of two chief kinds, the quill feathers and the clothing feathers. The most rudimentary of the latter are known as down. A *quill feather* consists of two principal parts: the quill, or calamus, and the vane, or vexillum. The quill is continuous with the central shaft, called the rachis, the two forming the stem of the feather. Projecting outward from the stem on each side are a large number of pointed and very flexible *barbs*. These barbs are located nearly at right angles to the quill and have extending from them at right angles smaller processes or *barbules*. These barbules hook together the barbs and give the web its form. In some feathers, as the hackle, and in the wing bar feathers of the male of some breeds, there is a portion of the upper outer edge of the feather not provided with barbules, which fact gives the feather its characteristic appearance. The down feathers are loose and fluffy. In this kind of feather the shaft is weak, and the barbs are not provided with barbules. The barbs like the shaft may be considered weak. These feathers give great warmth to the body. There are fiber feathers appearing as hair-like filaments, and called *filo-plumæ*. These are found scattered over the body; they are particularly abundant in the region of the head and the neck.

At each end of the quill is a small opening, or *umbilicus*. Inside the barrel of the feather there is *pulp*, which in young feathers is very vascular; the vessels entering by the proximal umbilicus are buried, along with part of the quill, in a papillated follicle of the skin. From this follicle the feather is developed. At the base of the shaft a secondary rudimentary quill is usually formed, which may be represented by a mere tuft of down. On the same general principle the smaller feathers are constructed. These cover the body, the upper parts of the legs, and the head, while the larger feathers and quills are confined to the wings and the tail. The longest quill feathers are those arising from the hand, called the *primaries*. Those arising from the forearm are called the *secondaries*. Those that are developed from the proximal part of the arm are called the *tertiaries*. The rudimentary pollux carries some feathers which form the *alula*, or bastard wing. The scapularies are feathers covering the scapula and the humerus. Covering the bases of the larger flight feathers are wing coverts consisting of several rows of small feathers. The quill feathers of the tail are called the *rectrices*. The rectrices have considerable mobility; their bases are covered by a row of tail coverts.

The pedal digits of the natatores are joined by a membrane, covered with scaly skin, which forms the web foot.

The feathers in many parts of the body are developed in rows, there being intervals, or elongated skin areas, between these groups of several rows, which are not provided with feather papillæ, but which are covered over by the feathers developed in front of these spaces. The definite feather lines, or areas, have been called *pterylæ*. The intervening tracts devoid of feathers are called the *apteria*.

The first outer covering of the bird, or baby chick, is a temporary one consisting of fasciculi of long filaments of down. These fasciculi on their first appearance, are enveloped in a sheath, which soon becomes ruptured and are entirely cast off by the time the baby chicks are ready to be taken from the nest or the incubator.

FIG. 80.—Photomicrograph of the skin of the neck of a S. C. White Leghorn hen. 1, The skin possessing an outer stratum, the stratum corneum, next the

stratum lucidum, next the stratum granulosum and underneath the stratum germinativum. Beneath the outer dark band representing the upper skin strata is the pars reticularis of the derma. 2, Papilla showing from inside to outside the hyaline feather wall, the stratum corneum, the Malpighian layer of the follicle, the corium. Inside the feather is noted the pulp-like material.

The down fasciculi, each emerging from its small quill, are succeeded by the feathers, which they apparently guide through the skin.

Feathers do not spring from all parts of the body alike; especially devoid of feathers are those parts where chafing and friction is greatest, as under the wings and in the groin.

At the end of the quill there is a small opening, the *inferior umbilicus*, into which projects a papilla of the dermis. Where the quill emerges from the skin there is another small opening, the *superior umbilicus*, from which springs frequently a small feather called the *hyporachis*. The shaft, or *rachis*, has a groove extending along that surface which lies next the body.

In the newly formed chick the first indication of feathers is the formation of papillæ, which is constituted by the upward growth of the dermis, or the sensitive and vascular parts of the skin. Then the skin immediately around the papilla sinks downward, so that later the papilla is inclosed in a follicle of the skin. The epidermis over the papilla is the same as over the rest of the surface. The horny outer layer of the epidermis forms for the growing feather a protective sheath which is cast off as the feather is formed. The feather proper develops from the underlying germinative layer, which as the feather develops, forms a cylinder of cells. The lower part of the cylinder is in touch with the papilla; this later becomes the quill (Fig. 80, No. 2). The upper part of the cylinder develops the web portion of the feather. As soon as the feather is fully developed the papilla, which has projected into the quill and nourished it, is withdrawn, and the quill becomes filled with a pithlike material forming septa, which extend in different directions.

Once a year, usually in the late summer or in the fall, the entire feather coat is changed. This process is called *molting*. During this time the bird appears in a somewhat depressed condition, the hen almost always ceases laying. Birds also molt in the spring, to a limited extent. The male at this time takes on the so-called breeding plumage, which is the most beautiful of the year. It has recently been established by Rice, that the young fowl, in reaching a stage of egg production, molts five times before the laying period begins.

The **structure of the skin** of fowls is similar to that of the skin of mammals. The skin consists of two layers the outer portion, or *epidermis*, and the inner true skin, the *cutis*, *corium*, or *dermis*. If we study a section of skin from the shank region of a fowl we find the outer portion is differentiated into two distinct regions, the rete Malpighii and the stratum corneum. The *stratum corneum* is the outer horny layer. The cells making up this portion are fusiform, flattened, and in regular rows. The nuclei in these layers of cells are not pronouncedly visible, and the outline of the cells not clear in a section such as used in our ordinary methods of study. The corneum is a compact mass of remnant cells which have lost the appearance and their texture of living cells. It thus becomes modified into scales upon the skin surface.

Between the stratum corneum and the stratum Malpighii there is another zone which consists of the cells undergoing a transitional stage from a cubical shape to a more flattened appearance, and gradually becoming more granular and hyaline.

The various parts of the epidermis are in close genetic relationship to one another. The upper layer of epithelium is constantly being desquamated. This casting off is compensated for by a continuous upward pushing of its lower elements. Cell proliferation occurs in the basal cells and in adjacent cellular strata of the stratum germinativum, or stratum Malpighii, where the elements are often seen in process of mitosis, or cell division. The young cells are gradually pushed upward. During their course they assume the general characteristics of the elements composing the layers through which they pass. This process is as follows: each cell changes first into a cell of the stratum Malpighii; then, when it commences the formation of keratohyalin, it changes into a cell of the stratum granulosum; later still, into a cell of the stratum lucidum; and finally into an element of the stratum corneum, where it gradually loses its nucleus, cornifies, and at last drops off.

The mesodermic portion of the skin consists of loose, subcutaneous connective tissue containing some fat. The amount of adipose tissue in the subcutaneous layer is subject to great variation; there are a few places in which there is little or no fat. The upper portion of this layer contains a few elastic fibers, which interlace and run in all directions. Numerous round or oval cells are found in the upper region. The lower and middle portions of the corium are richly supplied with blood-vessels terminating into capillaries, which penetrate the portions bordering the epidermis. Nerves giving off terminal branches also occur.

The various *colors* of the skins of fowls are due to the distribution of various quantities of two colors, orange-yellow and brownish-black.

The yellow pigment is probably carotin and xanthophyll, two pigments contained in association with the chlorophyll of plants, which the bird obtains in its feed. These coloring matters were formerly called lipochrome; but as lipochrome may be any coloring matter of fat, it is not sufficiently definite. This yellow pigment, when present, is diffused through all parts of the cell. When dilute, it gives a yellow hue; when concentrated, orange. It is found in the epidermis and in the fatty masses of and beneath the corium, and is probably identical with the yellow color of fat in other portions of the body and in the yolk of the egg. A yellow shank, in a heavily laying hen, soon loses a part of its pigment; this is also noted of the coloring matter in other parts of the body. This fact indicates that the coloring matter of the fatty part of the egg yolk is from the same source as that of the fat. The draught of this substance is more than normally and the reserve is being drawn upon. Feed rich in zanthophyll and carotin cause intense yellow colored yolks, and feeds poor in these substances cause the yolks to be a pale yellow. Cotton seed meal contains two pigments, one a yellow crystalline substance and the other a brownish resinous substance. Both of these pigments are probably deposited in the egg yolk. The eggs of some hens contain a large quantity giving a light brownish-yellow color to the yolk.

The brown or black-brown pigment is carried in microscopic pigment granules, which may be scattered through the ordinary cells or may be confined to special pigment cells. The former are confined to the epidermis; the latter may be found in both layers, but infrequently in the epidermis. When granules are present in the flattened cells of the corium, they occupy the nuclear region. They lie in short thin lines while those of the under portion of the Malpighian layer occur in oval groups. Where these granules occur in the rete layer, they cluster around the nuclei. In the colored skin there are dark pigment granules found in the corium and to a less extent in the rete layer. Hanau has described a definite cellular body which he found densely packed among the black-brown granules. There is a central body which sends out branches in all directions. In very dark skinned shanks these ramifying strands interlace, and form a compact network, which in many cases is so thick as to give the impression of a homogeneous mass. Here and there occur round or oval pigmented bodies, which Hanau concluded were the star-shaped cells with their pseudopod-like appendages contracted. Pigment cells commonly lie around blood-vessels, clearly indicating their course. They frequently form a compact tube, but more often are limited to fragments which only partly enclose the vessels. Pigment cells occur in several well-defined localities: in the upper portion of the cutis among the closely interwoven strands of connective tissue; in the region bordering the blood-vessels; in proximity to nerves, in nerve endings and in surrounding fat masses. Isolated granules are frequently scattered throughout the lower section of the corium. Barrows concludes that the lower bodies of pigment play little or no part in the color of the external shank, as they lie far below the opaque connective tissue. Melanin pigment granules are always contained in the pigment cells. When found in the Malpighian layer, the pigment cells are of an oval form.

Immediately below the epidermis in the shank skin, extends a space less in width than the row of columnar cells, which is devoid of pigment. The brown pigment is melanin, which in large quantities takes on a black hue.

White skin does not contain superficial pigment. Melanin has been observed in the study of white shank skin, but it lay at considerable depth or in quantities insufficient to be noticeable. In some breeds of fowls, as the Mottled Houdan, there are areas in which much melanin is irregularly deposited, which circumstance gives the leg its mottled appearance.

The yellow color is the result of a deposit of the yellow pigment in the fat of the shank. This may be deposited in both layers of the skin, or in the corium alone. When present it is diffused throughout the entire cell as well as throughout the intercellular substance. In young birds the Malpighian layer contains much yellow pigment. Old hens have only small quantities in the corneum. In breeds with normal yellow shanks old hens that have never laid eggs show a deep orange color in both the dermis and epidermis.

In blue shanked birds melanin is present only in the corium. The black pigment is seen through the semi-transparent Malpighian stratum, making it appear bluish instead of brown or black.

The black shank color results when melanin appears in the epidermis. Two forms of black pigment occur in the epidermis: granules in both layers and pigment cells in the rete Malpighii.

The green shank is produced where there is pigment in the epidermis and numerous melanin pigment cells in the upper corium. It is an optical effect due to melanin lying under the semi-transparent yellow epidermis. There is no melanin in the epidermis.

In the beak the corium of the skin is represented by a thin layer located between the **periosteum** and the stratum Malpighii. Numerous blood-vessels pass into it, and in the soft horn-like skin, occur sensory nerve fibers.

The nails originate from the epidermis, of which they are a modification. The nails of the toes are bent downward. They present a convex dorsal surface and are concave ventrally. The dorsal surface consists of a horny plate, which is set in a nail matrix. The ventral portion merges with the sides of the upper half, and, as the lower portion is the softer and wears faster, the nail has a sharp point and edge. The matrix is formed by a growth of the Malpighian layer of the cutis. A fold of skin lies over its posterior part. The epidermic cells of the dorsal, or nail, part, and the base of the ventral part grow fast. The outer cell layer gradually becomes horn-like.

The *spurs* are conical with a flat base. The basal part rests upon an enlargement of the shank bone. Soft structure is found between the horny spur and the bone. The upper cells, like those of the nails and of the skin, are constantly being worn or cast off, and new cells push up from the lower layers of cells. These newly formed flattened cells soon become cornified. The oldest formation is found at the tip and the youngest at the base.

EMBRYOLOGY OF THE CHICK

That a new individual may be brought into existence, there must be accomplished the union of the male element, or *spermatozoon*, with the female element, the *ovum*. This union is called *fertilization*. In the fowl this fertilization is accomplished at the anterior portion of the oviduct, after the calyx has ruptured and discharged its yolk, and before the albumen has been formed around it. The blastoderm is found on the surface of the yolk. One spermatozoon is all that is required; in fact, only one can be used in this union.

Spermatogenesis.—The spermatozoa are formed by the seminiferous tubules of the testis. From these cells, called the *spermatogonia*, are formed other cells called *spermatocytes*, which in turn form the *spermatids*, the immediate forerunners of the spermatozoa. During the period of multiplication the spermatogonia divide repeatedly by mitosis. Numbers of small cells are thus produced, each containing in its nucleus the number of chromosomes typical of the somatic cell of that fowl. In the second period the cells become larger and spermatocytes of the first order are formed. Then comes the period of maturation, during which two succeeding divisions rapidly occur. The first division results in the formation of two cells exactly alike. These are spermatocytes of the second order. They differ from the somatic cells in that they contain only one-half the typical number of chromosomes. The second division produces two similar spermatids from one spermatocyte of the second order. Therefore four spermatids exactly alike may be formed from one spermatocyte of the first order. From these spermatids the spermatozoa are formed (Fig. 55). The heads of the spermatozoa contain the nuclei derived from the spermatids; the necks contain the centrosomes; and the tail, consisting of three parts is probably formed from the protoplasm. Three parts of the tail are as follows: first, the pars conjunctionis, which unites the tail to the neck; second, the pars principalis, which constitutes the main length of the tail; and, third, the pars terminalis, which consists of an axial filament which transverses the entire tail and is surrounded by a protoplasmic sheath.

Oögenesis.—The ovum during its formation passes through three stages.

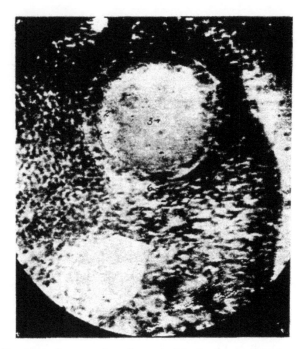

FIG. 81.—Section of ovum in a hen. 1. Nucleolus. 2. Nucleus. 3. Liquor folliculi. 4. Stratum granulosum. 5. Follicular cells. 6. Theca folliculi. 7. Peripheral stroma.

The first stage, that of *division*, takes place before the chick is hatched, and, according to Bradley, comes to an end about the time of hatching. This stage consists of the rapid formation of ova in the female chick. In the second stage, which begins about the time of hatching, there is an increase in the size of the units of the ovary, accompanied by *yolk formation*. At this time each ovum is in its own follicle (Fig. 81), and is surrounded by a layer of cuboidal cells and a *theca*. The theca is formed from the adjacent fibrous stroma. The third stage, that of *maturation*, commences during the development of the yolk and is complete after it has escaped to the oviduct. Maturation consists of each cell's dividing into two unequal parts. In each division the cell is split into a small cell known as the *polar body*, which is cast off and disappears, and a larger cell, which is the ovum proper. In this process half of the original chromosomes are cast off.

Fertilization.—The sperm travels rapidly; experiments have shown eggs to be fertile laid twenty-four hours after service by a male. When the ovum is discharged into the ovarian pocket it is surrounded by spermatozoa.

After the male pronucleus has united with the female pronucleus in the single-celled ovum, there is a cleavage of the cell in the long axis of the egg, making two cells; and then a cleavage at right angles, which progressively continues, makes the mulberry-like mass. The remaining content of the egg consists of food for the development of the embryo. From this mass of cells before the egg is laid the *blastoderm* is formed. The cells of the blastoderm are differentiated into two layers. The superficial layer is the *ectoderm* and the lower layer is the *entoderm*. In the newly laid egg the blastoderm may be observed. It is about 4 millimeters in diameter. It has a transparent central area, the zona pellucida, which is located over the subgerminal region. There is a peripheral, less transparent area called the zona opaca.

In the fertile egg, as soon as it is subjected to the proper temperature, cell multiplication in the blastoderm begins. The first signs of such change are noted in the pellucid area of the blastoderm where embryonal traces appear in the form of the parallel lines called the *plicæ primitivæ*, which diverge to form the cephalic dilatation. At about this time takes place the formation of the *myelencephalous columns*, in which the blood lakes expand in the surrounding halones and in the tracts along which pass colorless blood particles. These tracts extend from below the cephalic expansion to the peripheral sinuses, as the *proto-vertebræ*, which begin to appear at the sides of the myelon. The red color is acquired by the blood, and the heart by its movements, is made more manifest as the *punctum saliens*. A distinct membrane, the serous layer, is formed upon the germ and the blastoderm. The cephalic end of the embryo rises from the surface of the blastoderm, and then, curving down, sinks into it, forming for itself a kind of hood of the serous layer. This hood gradually extends from the margin of the fossa over the body, and, meeting a similar fold formed by the projecting and incurved tail, closes over the germ on the upper side, making a circumscribed cavity, which is the *amnion*. The progress of differentiation of layers of the blastoderm has, meantime, gone on beneath. The serous layer is in part reflected from the vascular and from the mucous layer. The mucous layer is concerned in the formation of the intestinal canal; and beyond this part, which is at first an open groove, the mucous layer expands over the yolk, which it ultimately incloses, the margins of the *vitellicle* so formed contracting and uniting at the side opposite the embryo at a sort of cicatrix, to which the last part of the abdominal yolk adheres. The vitellicle is richly vascular, and the surface next to the yolk is augmented by rugæ.

The fowl's egg, at about the fortieth hour, shows the buds from which the limbs are developed. A vesicle is seen to protrude near the anal end of the intestine, which, rapidly expanding, spreads over the embryo, acquiring a close adhesion to the amnion, but remaining distinct from the vitellicle, over which it spreads. It finally encloses the albumen and interposes itself between the latter and the lining membrane of the shell. Umbilical vessels are associated with this membrane. Hunter called this membrane the *allantois* from its containing urine, and Owen states that the sac which surrounds the albumen acts as the chorion or *placenta*; for it is most probable that from this surface the albumen is absorbed and the chick supported on its developmental food. The external part of the sac apparently acts as lungs as it comes into contact with the shell of the egg through pores of which there is an exchange of air. Oxygen is consumed and carbon dioxide is given off. The blood in the vessels of this membrane is in color more like arterial blood and that in the interior more like venous blood.

The embryonic mass of the incubating egg always floats to the top side. As the embryo grows it turns upon its left side, exhibiting a profile view; it then indents the yolk, and finally almost divides it into two parts.

The peripheral layers of cells rise from the margin of the germ mass, and extend and contract toward the opposite pole. This tract of germ substance is the *primitive streak*. Along the median line it next forms a furrow, which stops short of the ends of the streak. This streak terminates opposite the point from which the germ begins, and swells into the head. The median furrow expands upon it. The cephalic borders are next united by a thin layer of epithelial cells above the furrow, converting it into a cavity, or ventricle. The myelonal furrow is similarly covered by a layer, uniting the lateral columns. The embryonal trace becomes longer, narrower, and bends round the vitellus. A layer of epithelial cells forms a network over the whole dorsal surface of the embryo. Oblique striæ appear in the broadening germ mass radiating from the primitive streak. These indicate divisional segments. These beginnings of aponeurotic septa probably accompany and support nervous productions from the myelon columns.

Two transverse constrictions begin to divide the cephalic enlargements into three lobes, the second and the third of which expand into vesicles. An accumulation of cells at the side of the middle expansion appears to add greatly to its breadth. This forms the basis of the eyes.

The differentiation and the confluence of the cell constituents of the primitive streak have led to the formation of a pair of albuminous cords along the sides of the median furrow, forming the *myelon* proper. The cells exterior to and above them are converted into muscle and fibrous septa; and beneath the column is a jelly-filled cylinder, with a transversely striated sheath, pointed at both ends, forming the *notocord*. Its anterior point passes a little in advance of the acoustic vesicle. Beneath the notocord and surrounding the blastema is stretched the vegetative, or mucous, layer of cells, in contact with the yolk. Both the head and the tail of the now cylindrical embryo are liberated from the surface of the yolk. A fold of the blastema, reflected from the under part of the head, sinks like a pouch into the yolk, and soon includes the rudiment of the heart, like a bent cord, which begins to oscillate about the seventh day. From the midline of the inferior surface of the embryo, or its mucous layer, two longitudinal plates descend, diverging into the yolk-substance, and form the *primitive intestinal* groove.

The *ophthalmic vesicle* elongates and curves outward until the two ends almost come into contact. Between these two ends and beneath the delicate tegumentary layer connecting them the *crystalline lens* is formed. About the same time, the *otoliths* appear in the acoustic vesicles, which have now acquired a cartilaginous case. The *cerebral lobes* begin to be formed by a small fold, rising laterally and overlapping the forepart of the second enlargement, which has expanded to greater breadth. The *olfactory cavities* appear as small cutaneous follicles.

The two myelonal columns, expanding between the ear sacs and receding so as to show the notocord beneath, bend upward and inward, and unite, to be continued into the posterior of the *optic lobes*, thus commencing the cerebellar bridge across the epencephalic ventricle. The encephalic vacuities have begun to be filled by the granular basis of the cerebellar substance.

The *intestinal groove* begins to be converted into a canal at its two ends. Beneath the anterior end, and behind the heart, there gradually accumulates the cellular basis of the liver.

The commencement of the development of the *organ of hearing* is by a superficial depression of the cephalic blastema to meet the process from the encephalon, which forms the acoustic nerve. The lining of the depression becomes, on closure of the slit, the proper tunic of the labyrinth.

The vesicle of the labyrinth swells into four dilatations, of which three are ampullar and the fourth cochlear. The *ampullar dilatations* extend into very slender canals, at first almost in the same plane, by which they are brought into mutual communication. As the canals expand and elongate, they assume their characteristic relative positions as external, superior, and posterior, the posterior end of the external canal being extended beneath the posterior canal. The cochlear dilatation curves as it elongates. An inner layer becomes distinct from the common membrane and forms the *acoustic lamina*.

As in the development of the ear, so in the development of the eye, the production of the nerve process from the cerebral center is the first step; the infolding of the superficial blastema to meet the nerve is the next. The so-called *cutaneous follicle* becomes a circumscribed sac or vesicle, in which the changes and the development next proceed, converting the vesicle into an acoustic labyrinth or into an eyeball. In each case neural elements of two vertebræ become modified to lodge and to protect the sense organs, forming respectively the recess called *otocrane* and that called the *orbit*. The one is located between the occipital and the parietal vertebræ, and the other between the frontal and the nasal vertebræ. The part of the outer blastemal layer of the head which sinks to meet the process from the mesencephalic dilatation, rapidly changes its follicular into a vesicular state. The vesicle thus formed elongates, bending around the cell mass in which the *crystalline lens* is formed, and the meeting of the two ends results in forming the choroid fissure at the lower part of the eyeball.

The *mesencephalic process*, or optic nerve, expands at the posterior of the circular sac, and, in the course of mutation into eyeball, lines its posterior part with a layer which becomes the *retina*. The transparent layer covering the forepart of that sac and the inclosed lens is formed into the *cornea*. Other layers of the sac are formed into the *choroid*, the *ciliary processes*, the *iris*, and the *pecten*.

Of the appendages of the eye the membrana nictitans is the first to develop. Then develop the lower lid and, last, the upper lid.

After the development of the essential organs of sense, the skin is developed. Modifications of the skin form the outer ear and the eyelids. Then are formed the maxillary arch, the hyoidean arch, and the scapular arch.

The vessels which return the blood from the vitellicle are the transverse and the longitudinal *vitelline veins*. The first are so called because these trunks pass to the embryo at right angles to its axis. They are the largest returning canals. The *longitudinal veins* extend parallel with the axis of the embryo; they are of smaller size. The right anterior longitudinal vein becomes the right precaval and receives the remains of the right transverse vitelline vein, as the right vena azygos. The left anterior longitudinal vitelline vein is also persistent as the left precaval, and enters in the mature bird, as in the embryo, at the posterior or the lower part of the auricle. The left transverse vitelline vein is also subsequently reduced, by receiving only the vertebral veins of that side, to the condition of a so-called azygos vein. The main trunk of the post-caval is the result of the returning vessels from the abdominal viscera and the posterior limbs at a later stage of development. There is but one principal posterior longitudinal vitelline vein, and this anastomoses with the left transverse vein as it enters the embryo.

The *auricle*, which, by its dilatation of the left side, appears to be double, receives the venous blood at its right division. The left one, subsequently receives the veins from the lungs, is ultimately separated from the left precaval and the right auricle to which that vein is conducted and restricted.

The *ventricular part of the heart*, at the second day of incubation, is in the form of a bent tube, curving from behind downward, forward, to the right and upward, continuing insensibly into the part representing the aortic bulb, in which the septum first appears, and ultimately dividing the ventricle into two.

At this stage the piers of the *maxillary arch* appear as buds from beneath the eyeballs. The naso-premaxillary process is above their interspace. The piers of the mandibular arch and those of the hyoidean arch follow in close succession. The blastemal base of the scapular arch projects slightly at the sides of the fovea cardiaca; the piers, now separate, ultimately meet in front of the heart, and accompany it in its retrograde course. The mesencephalon is the largest of the segments of the brain which are connected with the eyeballs.

When the heart has assumed its form as such, distinct from the great trunks rising from it, the two *arteries* from the base of the ventricles appear. The artery to the right bifurcates, one division supplying the head and the wings, the other winding over the right bronchus. That to the left also bifurcates. Its left division arching over the left bronchus and anastomosing with the right arch a little below and behind the apex of the heart. Its right division arches over the back of the heart, bending to the right and anastomosing with the right aortic arch just above the outer ductus arteriosus. Each of these divisions of the left primary arterial trunks sends off a branch to its corresponding lung. As the lung expands, and especially at the beginning of the act of expansion toward the close of the period of incubation, the blood is diverted into the pulmonary vessels, and the channels below them shrink and disappear. The left primary artery is retained as the trunk of the pulmonaries, and, through the changes in the interior of the ventricle, this artery comes to discharge exclusively the ventricle corresponding to the right in mammals. The retained aorta rises from the left ventricle.

The *air-sacs* begin at the lower point of the lung, appearing like small hydatids, and extend further and further into the abdomen, in front of the kidneys. They are at first full of fluid. Soon after the development of the abdominal air-sacs others are developed.

The *lungs* are at first free, but afterward begin to be attached to the ribs and to the spine.

In the female embryo we first observe two *oviducts*, one on each side of the basis, or stroma of the ovarium, which appears in a relation to the primordial kidneys similar to that of the testes in the male. At the period when the permanent kidneys have sent their ureters to the cloaca, the oviducts have been developed as prolongations from that part, and, up to a certain point of development, they are of equal size and length. Subsequently the left oviduct alone continues to grow; the right remaining stationary or shrivels; occasionally it may be discerned as a rudimentary in the mature bird, but usually all trace of it has disappeared before hatching. The left oviduct expands above or at its free end into the infundibular orifice, where its parietes are very thin. As it descends, these increase in thickness, and the efferent tube gradually acquires the texture and form of an intestine. It is attached and supported by a duplicature of the peritoneum.

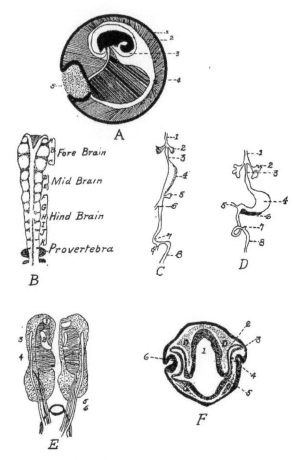

FIG. 82.—Embryological studies.

A. 1, The chorion and allantois. 2, The allantoic cavity. 3, The amnion. 4, The yolk sac. 5, A small quantity of remaining albumin.

B. Brain tube of a chick 25½ hours old showing partly closed brain tube with eleven folds of neuromeres.

C. The development of the alimentary tract. 1, Trachea. 2, Lung. 3, Esophagus. 4, Stomach. 5, Pancreas. 6, Bile duct. 7, V-shaped loop of midgut. 8, Cloaca. 9, Vitello-intestinal duct.

D. 1 to 6 inclusive, same as C. 7, Cæca. 8, Cloaca.

E. Kidneys, Wolffian bodies, and testes of an embryo chick. 1, The adrenals. 2, The genital. 3, Primitive oviduct. 4, Permanent kidney. 5, Ureters. 6, Duct of primitive kidney which conducts the excretion into the cloaca.

F. A transverse section of a chicks' head 48 hours of incubation. 1, Forebrain. 2, Pigmented layer of retina. 3, Ectoderm. 4, Nervous part of retina. 5, Optic stalk. 6, Invagination of ectoderm to form the lens rudiment.

At first, the right and the left ovaria are similar in size, but the symmetry is soon disturbed by concentration of development in the left ovary. The right ovary remains stationary and ultimately, in most birds, completely disappears by the time the chick is ready to emerge from the shell.

Three fetal membranes are developed, the *chorion*, the *amnion*, and the *allantois*. There is also developed the *yolk sac* (Figs. 82 and 83). The amnion is connected to the body wall at the umbilicus. The amniotic fluid is found in this sac. The chorion is at first surrounded by the albumen, but as the albumen is absorbed the chorion comes in contact with the inner shell membrane. It is probable that the chorion consists of ectoderm on the outside and of mesoblast on the inside. The amnion, on the other hand, is formed of mesoblast on the outside and ectoderm on the inner, or embryonal, side. The allantois springs from the embryo soon after the fourth day, and develops from the ventral wall of the primitive gut. By some embryologists, the yolk sac is included in the embryonic membranes. It commences as the splanchno-pleure surrounding the mass of yolk. It becomes smaller as the yolk is absorbed. At first its outline is round but later its walls become folded in. The yolk is dissolved and absorbed by the entodermic lining of the sac, and is carried to the embryo by veins called the vitelline veins (Fig. 83, No. *B*, 4), which ramify on the walls of the sac.

Some time after the fourteenth day, the chick assumes a position lengthwise within the egg shell so that the head is near the broad end of the egg. The head is bent upon the chest and the beak is usually tucked under the wing. Later the head assumes a position, by a double curve of the neck, so that the beak is in contact with the air cell. About the fifteenth day, the coils of intestine, which heretofore have been outside the abdominal cavity, are withdrawn into the abdominal cavity, as is also the abdominal yolk sac. As the chick pips out of the shell, the umbilicus becomes occluded.

The outer upper part of the tip of the beak is provided with a short, stout, spike-like arrangement, called the *egg tooth*. On the twentieth day this part of the beak is forced against the wall of the egg and gradually breaks through the egg shell. The breathing by the lungs commences some time before hatching; this is evidenced by the chick within the shell giving chirping sounds.

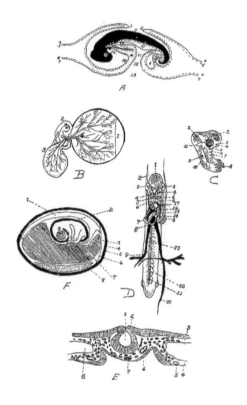

FIG. 83.

A. Longitudinal section of chick, 4 days incubation. 1, Cephalic fold. 2, Caudal fold. 3, True amniotic cavity. 4, Epiblast. 5, Somatic mesoblast. 6, Visceral mesoblast. 7, Hypoblast. 8, Future anua still closed. 9. The allantoic vessicle. 10, The mesentery. 11, Intestine. 12 and 13, Yolk sac. 14, Cavity of true amnion. 15, The mouth. 16, Pleuroperitoneal cavity. 17, Fore gut.

B. Membranes of the chick at the third day of incubation. 1, Membrane surrounding albumin. 2, Amnion. 3, Allantois. 4, Vitellicle.

C. The fore part of an embryo chick at the second day. 1, Mesencephalon. 2, The eye. 3, Olfactory organ. 4, Mandibular arch. 5, Maxillary arch. 6, Maxillary arch. 7, Hyoidean arch. 8, Scapular arch. 9, Depression organ of hearing. 11, Process from encephalon for union with nerve of hearing.

D. Primitive blood-vessels at second day of incubation (Owen). 1, Mesocephalon. 2, The right anterior longitudinal vein. 3, The eye. 4, The mandibular arch. 5, The hyoidean arch. 6, Scapular arch. 7, The same. 8, The ventricular portion of the heart. 9, The right transverse vitelline vein. 10, Posterior longitudinal vitelline vein. 11, Vertebral vein. 12, Tributaries of same. 14, The auricle. 16, Aortic bulb.

E., A transverse section of chick embryo, 29 hours incubation. 1, Neural canal. 2, Neural crest. 3, Somatopleure. 4, Splanchnopleure. 5, Omphalomesenteric vein. 6, Aorta. 7, Notocord. 8, Pleuroperitoneal cavity.

F. Chick embryo at the ninth day of incubation. 1, Allantois. 2, Amnion. 3, Air cell at large end of the egg. 4, Egg shell. 5, Outer shell membrane. 6, Inner shell membrane. 7, Yolk sac. 8, Albumin (Bradley).

The *circulation of the blood* is somewhat like the circulation in the fetus of quadrupeds. The primitive tubular heart (Fig. 83, No. *D*, 8) is bent in S-shape; the two ends are connected with blood-vessels; and later, by the development of the septa, the cavities of the adult heart are defined. The foramen ovale is found in the median septum. This opening brings the right and the left side into communication. The septa between the cavities of the heart are completed by the end of the sixth day. The two vitelline, or omphalo-mesenteric, veins carry the blood containing the nutrients from the yolk sac to the liver, where it is mixed with the blood drained from the intestines by the portal vein, and this blood is finally carried by the posterior vena cava into the right auricle of the heart. From here it passes through the foramen ovale to the left auricle; from the left auricle it enters the left ventricle, from which it is forced into the aorta, and then out into the systemic circulation. Practically all, if not all, of the blood of the pulmonary artery is sent into the aorta through a connection with this vessel called the ductus arteriosus.

Down appears about the thirteenth day of incubation. There are two kinds of down on the chick, one long, which comes first, about two or three days before hatching; a second, or fine, down forms at the roots of the other. As the embryo develops the air cell, at the large end of the egg which is developed between the two shell membranes, or *membranæ putaminæ*, gradually enlarges.

OUTLINE FOR LABORATORY STUDY OF THE CHICK

The objects of laboratory study of the embryos are as follows:

1. To study the living embryo.

2. To study the entire embryo:

With the dissecting microscope, as an opaque object.

With the compound microscope after killing, hardening, clarifying, and mounting.

3. To study embryos by dissection, in later stages.

4. To study the serial sections with the compound microscope.

THE LIVING EMBRYO

The egg is opened under warm physiological salt solution: 0.8 per cent. sodium chlorid in distilled water heated to a temperature of 38°C.

Gradually pick away the shell at the large end. Note that there are two membranes and an air cell. Strip off the membranes. When sufficient shell and membranes have been removed from the large end, invert the open end of the egg in the salt solution and allow the contents to flow out. Care must be taken not to break the yolk. The embryo, or blastoderm, lies upon the surface of the yolk, which is usually turned with this body uppermost. Separate the blastoderm by cutting around the outside of the area vasculosa. In doing this a small pair of slightly curved scissors is needed. After the embryo, or blastoderm, has been separated, gently float it into a watch crystal with the flat bottom submerged in the salt solution. The watch crystal with its contents may now be gently lifted out. Next remove the vitelline membrane. The vitelline membrane is the delicate transparent membrane covering the blastoderm, The embryo is now ready for study.

THE PREPARATION FOR STUDY OF ENTIRE EMBRYOS AND SECTIONS

The following processes may be used in killing embryos up to four days, or ninety-six hours, of age. After removing the embryo as described above, spread the blastoderm out in the watch crystal and pipette off the salt solution. Allow it to stand till the edge of the tissue begin slightly to adhere. Then slowly add the killing fluid by aid of a pipette, dropping it on the center of the embryo. The pipette must be held low or the mechanical interference will dislocate the parts.

Older embryos are submerged with their membranes intact into the killing fluid. The quantity of fluid should be several times the bulk of the specimen. Kleinenberg's picrosulphuric acid may be used as a killing fluid. This fluid is a saturated solution of picric acid plus 2 per cent. sulphuric acid, to which is added twice its volume of water.

Chick embryos from one to two days old should be left in this fluid from one and one-half to six hours. Embryos from two to four days old two and one-half to six hours. Remove the specimen from the killing fluid, and place it in 70 per cent. alcohol. Change the alcohol every twenty-four hours until the color ceases to come out of the embryo. Preserve in 80 per cent. alcohol.

If the specimen is to be mounted whole, transfer it from 80 per cent., then to 50 per cent., then to 35 per cent., and, finally to water. Small embryos should remain in each fluid thirty minutes and large ones sixty minutes.

The following method may be used for *staining embryos*: Dilute Delafield's hematoxylin with four times its volume of water. To every 6 cubic centimeters of this diluted hematoxylin, add one drop of Kleinenberg's undiluted picrosulphuric acid, and leave specimen in the fluid thus prepared until it is stained through. This will require from one to three hours. Now pass up through the series of alcohols to 70 per cent. Next extract the excessive stain with 1 per cent. hydrochloric acid in 70 per cent. alcohol. Wash repeatedly with 70 per cent. alcohol to free from the acid; then transfer the specimen to 80 per cent. alcohol and leave in this for several hours for complete removal of the acid; then transfer to 95 per cent. alcohol for thirty minutes. Allow the specimen to remain in absolute alcohol for one hour. Introduce a layer of oil of cloves or xylol beneath the alcohol. This may be done by gradually allowing the fluid to run down the side of the bottle. After the embryos have sunk into the oil and begun to appear transparent, remove the fluid and add fresh oil. After the specimen is sufficiently transparent, mount in balsam, supporting the cover slip so that it will not rest on the embryo.

In staining for section, place the embryo in borax carmine from the 50 per cent. alcohol and leave twelve hours. Then wash in 50 per cent. alcohol; after which transfer to 70 per cent. alcohol for six hours. Clarify in oil of cedar or oil of cloves, and place in melted paraffin for two hours. Imbed in paraffin, and section.

Embryos may be hardened, imbedded, and sectioned after the usual methods, using either paraffin or celloidin, and the sections stained with hematoxylin and eosin after sectioning.

POINTS TO BE OBSERVED IN THE STUDY

1. THE EMBRYO TWENTY-NINE TO THIRTY-FOUR HOURS OLD WITH FROM TEN TO FOURTEEN SOMITES.

A study of the egg.

In opening the egg, observe that the shell membranes are double.

Observe that the shell is porous.

Note the air cell at the large end of the egg and note that the space or cell lies between the outer and the inner shell membrane.

After the egg contents have dropped into the salt solution, note extending from the ends of the yolk the twisted denser cords of albumen. These are the chalazæ, which act as stays to the yolk. Note that the yolk is surrounded by a delicate membrane. This is the vitelline membrane. The yolk is the true ovum and serves as food for the developing embryo.

A study of the living embryo.

Note the amount of yolk that is covered by the blastoderm. Note the slipper-shaped, transparent center of the blastoderm. This is the area pellucida. In the center of this there is a narrow white streak. The area opaca is the area lying external to the area pellucida. In this there is the area vasculosa.

A study of the embryo entire, including the vascular area.

The blood islands show as irregular deeply stained masses in the vascular area. At this stage they are inclosed in wide anastomosing tubes, the extra-embryonic blood-vessels, which open peripherally into the bounding sinus terminalis.

The following structures may be identified:

The neural tube forming the axis of the embryo. In the anterior region may be noted the forebrain, the optic vesicles, the midbrain. The hindbrain is subdivided into the neuromeres. The cord, or myelon, of the neural tube back of the hindbrain is closed in front but is open behind.

The head projects above the blastoderm. The fold which unites the ventral surface of the head with the blastoderm is called the head fold.

The diverging folds of the myelon encloses the primitive streak.

On each side of the neural tube are the mesoblastic somites. The series is continued behind by the undivided segmental plate.

The heart is beneath the hindbrain. The portion of the body cavity in which it lies, is bounded in front by the head fold and behind by the diverging limbs of the splanchno-pleure. Its posterior or venous end receives the vitelline veins from the vascular area. The anterior or arterial end is prolonged into the ventral aorta.

Note that the axis is somewhat bent.

The head fold of the amnion is noted to extend over the anterior end of the head.

A study of the transverse sections.

The sections about 20 micro-millimeters thick should be cut serially, and so mounted. Knowing how many micro-millimeters the embryo is in length will enable one to make a diagram of the fetal structure in the study of the series of transverse sections. The sections should be drawn in the order studied so as to obtain relative structural ideas.

With the microscope study the following regions:

1. Optic vesicle.

2. Midbrain.

3. Posterior half of the heart.

4. Myelon in the closed region.

5. Myelon in the open region.

6. Through the primitive streak.

2. THE EMBRYO TWENTY TO TWENTY-FOUR HOURS OLD WITH FROM TWO TO SIX SOMITES.

Compare the parts in this embryo with those of the embryos from twenty-nine to thirty-four hours old. Note and describe the relations of the embryo, the area pellucida, and the area opaca. Note how much of the yolk the blastoderm covers.

In studying the entire embryos, note the condition of the medullary plate. Observe if the tube is formed in any part; how far back the head plate extends, if the heart can be seen; and if the primitive streak is longer or shorter than in 1.

Study relations of structures and make drawings through the point of divergence of the walls of the fore gut; also through the somatic region, and through the primitive streak.

Study all parts, as the head fold, the heart, and the fore gut.

3. THE EMBRYO FORTY-FOUR TO FORTY-EIGHT HOURS OLD WITH TWENTY-FOUR TO TWENTY-NINE SOMITES.

Remove the embryo with the entire area, and preserve it.

Note and carefully describe the changes visible to the naked eye since the thirty-fourth hour.

In making a study of the entire embryo we note there has been a rapid growth of the dorsal surface of the head, which has become more bent. This bend, in the region of the midbrain is called the cephalic flexure. The forebrain and part of the midbrain form almost a right angle with the rest of the head. The head is compressed laterally and free from the blastoderm. The dorsal side of the trunk is turned up, and there is a twisting of the axis of the embryo just back of the heart. The tail fold begins at about this time and may or may not be visible. The optic vesicles are relatively smaller in relation to the brain than in 1. Note the part of the forebrain to which they are attached. Observe the inner and outer layers of the retina, the lens, the choroid fissure, and the cavity of the vitreous humor.

Note the auditory vesicles and whether or not they are closed sacs.

Note that the heart has grown in length and has become doubled on itself. The two ends are fixed. Note the relation of the heart to the afferent and efferent blood-vessels. Note that two, and possibly three, visceral pounches are visible. Note whether or not they are ventral to the midbrain. The first, the hyomandibular pouch, is bounded in front by the first visceral, or mandibular, and behind by the second visceral, arch. The second pouch is bound in front by the hyoid, and behind by the third visceral arch. The third pouch is bounded in front by the third visceral arch, and behind by the fourth. Note the number of mesoblastic somites and the condition of the mesoblastic segmental plates.

Note how far back the foregut is closed. Locate the head fold of the amnion, and note how far back it is closed. Note changes that have taken place in the vascular area. In studying the sections it will be found that a section cut transversely to the trunk will pass horizontally through the forebrain and through the midbrain.

Study a section through the trunk a short distance behind the heart. Observe the elevation of the axis of the body. Note the way in which the lateral folds, or the lateral limiting sulci, in the somatopleure, delimit the embryonic from the extra-embryonic area.

Note the appearance of the mesenchyme, the approximation of the two dorsal aortæ, the appearance of the amniotic folds. Observe in the mesoderm the posterior cardiac veins, and the myotomes, or muscle plates. The sclerotome is made up of the mass of mesenchyme between the myotome, on the one hand, and the neural tube and the notochord, on the other. Note the folding of the splanchno-pleure, and note if there is present the Wolffian duct, or the nephrotome.

Study sections through the optic vesicles. Note if there is the beginning of the lens. Do you note the diverticula of the pharynx? Can you identify the closed amnion and the chorion?

Study sections through the auditory pit. Note fusion of the gill pouches with the ectoderm. Note the blood-vessels. Study sections through the region of the heart, through the roots of the vitelline veins, and through the primitive streak, if it is still present.

For this study it will appear that the anterior end has developed in advance of the posterior end. The tail fold has probably just begun.

Write a description of the pharynx, and of the circulation at this stage.

4. THE EMBRYO SIXTY-EIGHT TO SEVENTY-TWO HOURS OLD WITH CERVICAL FLEXURES FORMED.

In a study of the living embryo note the changes visible to the naked eye since forty-eight hours old. Note the difference in the blood-vessels of the vascular area. Name the arteries and veins. Note the beating of the heart.

In a study of the entire mount note that a second, the cervical flexure, has appeared in the head. Note that the tail fold is well formed. Note the position of the embryo on the blastoderm. Determine if the amnion is completely closed. Note the olfactory pits on the ventral surface of the head, a short distance in front of the optic stalks. Note the telencephalon, a rudiment of the cerebral hemispheres and an extension of the primary forebrain. It is bilobed anteriorly. The optic stalks are attached to the floor near the anterior end of the thalamencephalon. Note the infundibular region, which is the depressed region behind the optic stalks. In the roof of the thalamencephalon there is a short diverticulum, the epiphysis. The mesencephalon, or midbrain, forms the apex of the cranial flexure, and is united to the hindbrain by a narrow isthmus. The metencephalon, or rudimentary cerebellum, appears as a thick portion on the most anterior division of the hindbrain. The rest of the hindbrain is provided with a transparent roof and constitutes the myelencephalon, or the rudimentary medulla oblongata. Observe the inner and the outer wall of the optic cup, the lens, the choroid fissure, and the posterior, or vitreous, chamber.

Note the form of the otocyst, or auditory sac. Note above which visceral arch it lies. Note the number of visceral clefts. The visceral arches are formed by the thickening of the walls of the bounding clefts. The visceral arches are as follows: the first is the mandibular, or hyomandibular, arch, which is in front of the first cleft. From this there is developed the lower jaw. Note if there is a maxillary process arising from the dorsal angle of the arch. The second arch is the hyoid arch, which is located behind the first cleft. Then follow in order the third, the fourth, and the fifth visceral arches. Note above the mandibular arch the rudimentary trigeminal ganglion, and above the hyoid arch the rudimentary acoustico-facialis. The latter is in contact with the anterior walls of the auditory sac. The rudimentary glosso-pharyngeal ganglion is noted above the third visceral arch. The vagus, or pneumogastric, ganglion is located above the fourth and the fifth. Note the form and the position of the heart. Note the anterior and the posterior limb rudiment in the trunk.

In a study of the sections it is found that cuts transverse to the trunk pass about horizontally through the forebrain. At this age the following sections of the embryo should be studied:

First, through the hindbrain, at which level will be noted the auditory sacs, the neuromeres, the trigeminal, the acoustico-facialis, and the glosso-pharyngeal and the vagus ganglion.

Second, through the upper part of the pharynx, at which level will be observed the midbrain, the hindbrain, the visceral pouches, the nerves, and the blood-vessels.

Third, through the choroid fissure of the optic cups. Note the parts of the eye, and, on the other side of the section, the heart.

Fourth, a study of a section through the olfactory pits.

Fifth, a study of a section through the pancreatic and the hepatic diverticula.

Sixth, a study at the beginning of the allantois through the hind-gut.

At this age it is of interest to study the systems of organs. Observe the manner in which the splanchno-pleure folds to form the walls of the intestine. Note the commencement of the mesentery. Note that the foremost part of the alimentary tract is formed from the stomodeal invagination of the ectoderm. The hypophysis is formed from a dorsal outgrowth of this. Note its relation to the brain.

The following structures are formed from outgrowths of the ectoderm at this stage:

First, the visceral pouches.

Second, the median rudiment of the thyroid. This is an outgrowth from the pharynx between the two hyoid arches.

Third, the rudimentary lungs, which develop in a pair from a median ventral diverticulum of the alimentary tract, just behind the last visceral pouch. The esophagus is just posterior to this. The esophagus, very short at this stage, is continuous with a slightly wider part that develops into the stomach.

Fourth, the first liver diverticulum, and, at a short distance posterior to this, the second liver diverticulum.

Fifth, the pancreas is at a point where the intestine opens ventrally. It first appears as a slight thickening of the dorsal angle of the intestine.

Sixth, the ventral wall of the hind-gut forms a wide evagination.

Seventh, the beginning of the allantois.

After the whole series of transverse sections have been studied and drawn, construct a longitudinal section of the fetus, including a reconstruction of the alimentary tract.

At this stage the **heart** is a simple tube. The following divisions are distinctly visible: the auricular portion, the ventricular portion, the sinus venosus, and the bulbus arteriosus. The union of the two ductus Cuvieri and the ductus venosus form the sinus venosus. The two ductus Cuvieri are formed by the union of the anterior and the posterior cardinal veins. The ductus venosus is formed by the union of the small right and the large left vitelline vein. These latter veins return the blood from the yolk sac. The sinus venosus empties into the single auricle above which it is located. The single auricle is later divided into two chambers, the right and the left auricle. At this stage it is widest in the lateral direction. The auricle empties directly into the ventricle. The ventricle lies ventrally and behind the auricle. This location is due to the bending of the heart at this stage of development. Its hindmost portion forms the future apex of the heart. If the series of sections be studied from the posterior forward, the ventricle will be first to appear in the sections. Just beneath the auricular portion of the heart there is the bulbus arteriosus. The bulbus arteriosus soon divides into a number of aortic arches. There is an ascending pair in each of the visceral arches. The dorsal aorta is formed by the union of the aortic arches above the visceral arches. The aortic arches are first continued a short distance forward as the carotid arteries. The dorsal, or posterior, aorta passes backward under the notochord. The dorsal aorta divides into two parallel aortæ which give off on each side the vitelline arteries. Note other branches of this aorta.

The veins at this stage consist of the anterior and posterior cardinal, the ductus venosus, the ductus Cuvieri, and the vitelline veins.

Make drawing of the circulatory system after a completion of the study of the series of sections.

Make a study of the **nervous system** according to hints already given.

The dorsal and the ventral roots of the spinal nerves are given off separately, and secondarily unite. From the neuroblasts of the cord there are at regular intervals outgrowths representing the ventral roots. From the neural crest there develop segmental collections of neuroblasts which form the spinal ganglia, from which the dorsal spinal nerve roots develop. In fact this developmental stage can be observed in embryos only forty-eight hours old, first appearing as a line of cells springing on each side from the angle between the neural canal tube and the external epiblast. In the section from the embryo seventy-two hours old there are observed the rudiments of the development of these spinal nerves. Four primary ganglia develop in the neural crest of the head. These ganglia are as follows: the acoustico-facialis ganglia, which is located over the hyoid arch; the ganglia of the trigeminus, which is located over the mandibular arch; the ganglia of the glosso-pharyngeus, which is located over the third visceral arch; and the ganglia of the vagus, which is located over the third and the fourth visceral cleft.

In this stage of development the trigeminal and the acoustico-facialis are clearly visible.

On each side and dorsal to the aorta is noted the Wolffian body, or mesonephros. The Wolffian body (Fig. 82, *E*) consists of a series of tubules imbedded in the mesenchyme. The openings into the Wolffian duct lie just beneath the cardinal vein.

In a study of the **Wolffian duct** determine just how far anteriorly and how far posteriorly it extends. Note whether it empties into the cloaca.

Each tube beginning in a blind extremity is later dilated. It has a thin wall, and is situated near the median portion of the Wolffian body. The tubule proper passing transversely, opens into the duct. The upper wall of the thin walled part is invaginated by a mass of mesenchyme that receives a small vessel from the dorsal aorta. The Malpighian corpuscle, consisting of a glomerule and Bowman's capsule, is thus established. In the study of the four days old chick note the further development of these parts.

In a study of a chick four days, or ninety-six hours old, note to what extent the yolk is covered by the blastoderm. It will be noted that the embryo lies in the extra-embryonic cavity. This cavity is bounded above by the chorion and below by the splanchno-pleure. In removing the amnion from the embryo, note the relation to the somatic umbilicus. Note the relation of the splanchnic umbilicus to the splanchno-pleure. What relation has the allantois to the above?

In examining the head, locate the cerebral hemispheres, and note their development. Locate the pineal gland. Note changes in the olfactory pit and the eye. Locate the lens and the choroid fissure. Note that the maxillary process of the mandibular arch lies beneath the eye and behind the olfactory pits. Note the otocyst and the relations of the other arches to the above structures.

In a study of the **trunk**, note the tail, the allantois, Wolffian ridges, the heart, and the condition and the position of the rudimentary limbs.

Make drawing of embryo from the side view. Carefully cut off the head immediately behind the last visceral arch, and study and draw the structures observed on the ventral side. Note the maxillary processes, the mandibular and the hyoid arch, the nasal pits, and the fronto-nasal process, which is just beginning its development.

In a study of transverse sections observe from your drawing at what level the section is made. Study and draw a section made through the region of the anterior limbs. Note the spinal ganglion, the muscle plate, or myotome, the condensation of mesenchyme around the notochord, the pancreas, the liver, and the intestine; and note the distribution of the mesenchyme. Note the ventral roots of the spinal nerves, and the neuroblasts in the spinal cord.

In sectioning the embryo from before backward, the first sections will pass horizontally through the hindbrain and the midbrain region. Note the parts, including the auditory vesicle. In the first series of sections also locate the ganglia of the pneumogastric, or vagus, the acoustico-facialis, the trigeminus, and the glosso-pharyngeal nerves. Note the notochord and the cardinal veins. After the disappearance of the ear, we observe the midbrain, which, is located at one end of the section, and the cord at the other end. The region between lies just above the pharynx. In the following series study the visceral arches. Note the third pair of cranial, or motor ocular, nerves. This latter nerve springs from the floor of the midbrain. From the ventral prolongation of the floor of the thalamencephalon there arises the infundibulum. The hypophysis is located just beneath the infundibulum. At this stage the hypophysis appears as a tube to empty into the mouth. This is an ingrowth of the oral epithelium.

In a study of a section through the center of the eye we should observe the lens and optic stalk. This is in the region of the optic chiasm. Note the choroid fissure and the pineal gland, the latter appearing just beyond the eyes. Just forward lies the telencephalon, or rudiments of the cerebral hemispheres.

In a study of the **alimentary tract** we note that the mouth is bounded by the mandibular arches. Note the maxillary processes, and the ventral surface of the head. A finger-like diverticulum, extending from the roof of the ruptured double membrane, formerly separating the pharynx from the arches, forms the hypophysis. The great development of the visceral pouches makes the pharynx rather complex. In studying sections horizontally through the pharyngeal region note the various visceral arches and pouches. Note the arteries of the thyroid diverticulum.

In the series note the changed development of the lung rudiments, the glottis, the esophagus, the trachea, and the bronchi. The bronchi appear in pairs. Note that the liver has assumed proportions, and that it surrounds the common trunk of the vitelline veins, which it divides into two parts. The sinus venosus lies close to the heart. The ductus venosus is also surrounded by the liver. Above the tip of the ventricle we note a dilatation which represents the stomach. The hepatic, or bile duct is located immediately behind the stomach. This duct is formed by the fusion of the right and the left duct. Locate, draw, and describe the pancreas; trace the intestine; locate the splanchnic umbilicus, or yolk stalk; locate the allantois stalk, and trace its connection with the hind-gut.

In the series, locate and study the Wolffian ducts, the beginning of the Müllerian duct, the embryonic kidney, or mesonephros, the permanent kidney, or metanephros. The urino-genital ridge is made up of all the above except the last named. The urino-genital ridge forms a rounded projection on each side of the mesentery into the dorsal angles of the body cavity.

The Wolffian ducts empty into the cloaca. There are two ducts which may be traced far forward, and which are found to extend backward along the lateral margin of the ridge to the cloaca. Along the greater part of their length we note tubules emptying into them.

Beginning near the anterior end of the urino-genital ridge, we note that the Müllerian ducts arise from a thickened line of epithelium. The greater part of the ridge is formed by the mesonephros. This is made up of a series of tubules in each of which we may distinguish two parts as follows: a tuft of capillaries from the aorta forming the glomerules, surrounded by a thin walled invaginated capsule, making up the Malpighian corpuscle; and the tubules proper. The tubules lead from the corpuscles, or glomerules, to the Wolffian duct.

The germinal epithelium constitutes the essential portions of the gonad, or ovary, or testis. The germinal epithelium arises from a thickening of the peritoneum of the median wall of the ridge. The gonad is found near the anterior end of the ridge. At about this age of the embryo there should appear the primitive ovary or testis. Near the posterior termination of the Wolffian duct and from the dorsal diverticulum there arises the ureter, or metanephros duct.

In these series there should be studied the heart and circulation, which will be found similar to the sections from the embryo seventy-two hours old.

Work out a summary of the relations of the allantois and the yolk stalk to the intestines; the relations of the Wolffian ducts to the intestines; the origin of the ureters from the Wolffian ducts; the relations of the lungs, the liver, and the thyroid gland; the relations of the blood-vessels; a study of the muscle plates; the relations of epiphysis, hypophysis, infundibulum, mouth, and pharynx; the relations of the vagus, or pneumogastric, trifacial, acoustico-facialis, and glosso-pharyngeal nerves to the visceral arches.

In addition to the study of embryos at the end of each of the first four days of incubation, a study should be made of preserved museum specimens each of which represents the development of a day up to and including the twenty-first day.

THE DERIVATIVES OF THE GERM-LAYERS

From the three primary germ-layers are developed the various tissues and organs of the body by metamorphoses which may be referred to the two fundamental processes of specialization, or the adaptation of structure to function, and of unequal growth, which latter results in the formation of folds, ridges, and constrictions.

From the ectoderm are produced:

The epidermis and its appendages, including the nails, the epithelium in connection to the feathers and the feathers.

The infoldings of the epidermis, including the epithelium of the mouth, epithelium of the salivary glands and the anterior lobe of the pituitary body, or hypophysis.

The epithelium of the nasal tract with its glands and communicating cavities.

The epithelium lining the external auditory canal, including the outer stratum of the membrana tympani.

The lining of the anus.

The epithelium of the conjunctiva and of the anterior part of the cornea, the crystalline lens.

The spinal cord, the brain with its outgrowths, including the optic nerve, the retina, and the posterior lobe of the pituitary body.

The epithelium of the inner ear.

From the entoderm are produced:

The epithelium of the respiratory tract.

The epithelium of the digestive tract, from the back part of the pharynx to the anus, including its associated glands, the liver, and the pancreas.

The epithelial parts of the middle ear and of the eustachian tube.

The epithelium of the thymus and the thyroid bodies.

From the mesoderm are developed:

Connective tissue in all its modified forms, such as bone, cartilage, lymph, blood, fibrous and areolar tissue.

Muscle tissue.

All endothelial cells, as of joint-cavities, bursal sacs, lymph sacs, blood-vessels, pericardium, and endocardium, pleura, and peritoneum.

The spleen.

The kidneys and ureters.

The testicles and the system of excretory ducts.

The ovary and oviduct.

PREPARATION OF STRUCTURES FOR STUDY

It is hoped that the following suggestions will be helpful in the laboratory work.

Directions for Dissecting Muscles.—The muscles that are brought into great play in movements of the bird's limbs are dark carmine in color, while those which are not brought greatly into use are pale or white in color.

The tendons are made up of white fibrous connective tissue, are very dense, and pearly white in color.

In securing a bird for dissection of muscles it is best to select one in rather poor flesh, as the fat is annoying. The bird may be chloroformed or killed in a bell jar by aid of illuminating gas. After the bird is dead pluck all the feathers and immerse it in a 10 per cent. solution of formaldehyde or of 80 per cent. alcohol. It is best to puncture the abdominal wall so that the fluid may at once fill the abdominal cavity and more readily gain access to the chest cavity; and to puncture the skin at various points so that the liquid may more quickly become disseminated among the muscular structures. Post-mortem changes quickly take place if these precautions are not taken. The liquid surrounding the carcass should be at least twice the quantity of the bulk of the carcass.

The first dissection should be to lay bare the dermal muscles. The dermal muscles are of two kinds, true dermal and dermo-osseous. The dermal muscles have their origin and insertion in the skin, and control the movements of the different groups of feathers. The dermo-osseous have their origin on some part of the skeleton, and insert to the integuments.

The dermal muscles vary with the characteristics of the bird, we do not find all the known dermal muscles in any one specimen. A cock of the Cornish breed will show these muscles best developed. Birds possess an enormous system of minute muscles divided up into an infinite number of fasciculi, to act harmoniously upon the feather quills and to agitate collectively the plumage. By the aid of a low-power lens the action of the feather muscles in the large quill butts of the wing or the tail may be studied.

The muscles may be studied in groups as outlined in the text. Make an incision through the skin down to the bone on the superior part of the head, parallel and close to the base of the upper mandible, and extending completely across. From the outer end of this make an incision backward and down to the skull and posteriorly. The muscles of the upper part of the neck will then be exposed. The straight incision should extend to about a half inch on the inside of the upper eyelid of the same side. Reflect the flap of skin from the top of the skull, and carefully examine the under side of it in the median line, where it overlies the frontal region. The dermo-frontalis will be observed if it be present. In many birds, especially in females, it may not be discernible, and may be considered absent. To expose the circumconcha make an incision completely around the ear; then carefully dissect to the ear base. A dermal circular muscle should be observed. To expose the dermo-temporalis extend the longitudinal incision down the back of the neck to a point between the clavicular heads, carrying it just through the skin and about one-fourth of an inch to the side of the median line. Remove the skin from the throat and the anterior portion of the chest. Lay open the alar and parapatagial duplicatures of the skin. This exposes a number of dermal muscles. The dermo-temporalis is now observed to extend from a small depression just above and anterior to the temporal fossa. It makes slight attachments to the temporal muscle, which it covers, and extends backward as a thin ribbon-shaped muscle, the fibers blending with those of the cleido-trachealis, and becomes lost upon the skin in front and opposite the shoulder-joint. At times its fibers blend with those of the dermo-tensor patagii.

As the musculature of the fowl is loosely arranged, the rest of the dissection is easily done if care be exercised.

The ligaments may be dissected after the completion of the study of muscles, using the same subject or, a two pound broiler be prepared by killing in the gas chamber, plucking the feathers and parboiling just till the flesh becomes tender and is easily removed, it will be observed that all structures can be removed from the points exposing the ligaments distinctly. The ligaments appear swollen and more easily observed for study.

Directions for the Study of the Viscera.—Carefully remove the right and the left abdominal and thoracic walls, allowing a strip of tissue to remain in the median line to hold the organs in their normal position. To open these cavities it is necessary to use the bone saw and the scalpel. The organs may now be studied from each side. To make a longitudinal section through the median line, select a small bird, one weighing not more than 2 pounds, kill, and preserve in a 10 per cent. solution of formaldehyde for three days. Then with a sharp, thin, long-bladed knife make an incision at one sweep through the median line of the body down to the back bone, and with the bone saw section through the vertebræ. If it is difficult to cut through the breast-bone, saw through before making the incision. In small birds the entire cut may be made without the aid of the saw.

Directions for the Study of Arteries.—Arteries should be injected. Veins are usually more or less filled with blood so that the tracing of these is not so difficult as the tracing of uninjected arteries. Nerves are white and no difficulty is usually encountered in tracing them.

The courses of arteries, veins, and nerves are side by side, and many of them, as in mammals, are arranged in the order of veins, arteries, nerves, the veins being in front.

The injection apparatus consists of the following parts: air-compression chamber, to which is attached a pressure pump. A manometer made of glass tubing 6 millimeters in diameter, inside measurement. This tube is partly filled with mercury, and a scale in centimeters is made from the top of the right-hand tube (Fig. 84, No. 8). This ruling, or gauge, should be about 15 centimeters long. Extending from the right extremity of the U-tube is a small rubber tubing which is attached to the chamber containing the injection fluid; and extending from the inferior part of this chamber is another tubing which has the injection needle attached to the free end.

The injection should be done under 120 millimeters pressure. The stop cock of the pressure chamber is released sufficiently to raise the mercury in the U-tube six centimeters, which multiplied by 2, the amount of work required to raise two columns, makes 120 millimeters pressure. If this pressure be maintained, all vessels should be injected without rupture. The same process may be used in injecting the air cells through the trachea.

The injecting material may consist of one part finely sifted plaster of Paris, four parts water, and sufficient gentian violet to make a violet color. For the coloring, red aniline may be used in preparing this material. The dye should be dissolved in the water to be used in making the injection liquid. Caution must be used and the work rapidly done, as the plaster soon sets, or becomes solid, in the needle or in the tubing. A small cannula should be used, since the endothelial lining of the arteries are easily injured and difficulty may thus be created.

Select for arterial dissection an old cock, as in a bird of this kind the arteries are larger and the difficulties are reduced. Select for bleeding and injection the ischiadic artery in the thigh region. With the sharp point of the thin blade of a knife make an incision lengthwise of the artery being careful not to strip back the endothelial lining of the artery. Allow as much of the blood as will escape before injecting; in fact, the arteries should be thoroughly emptied, so that there is no longer danger of a clot's plugging some vessel and thus preventing its filling. Since the blood of most birds coagulates in about thirty seconds, this work must be done rapidly, care being exercised to keep the flow running as long as possible. After bleeding is completed, insert the cannula and tie the vessel tightly around the cannula to prevent the escape of the injecting fluid. See that all connections are sufficiently tight to prevent the escape of liquid under pressure. After the injection is completed, remove the cannula and tie the artery with a small twine, preferably cotton. Quickly remove all injecting fluid from the needle, the tubing, and the injecting chamber.

During the operation of bleeding the cock may be chloroformed, care being taken not to administer chloroform to kill him; for it is necessary to maintain life as long as possible so that the heart may be kept beating and all blood possible drained from the arteries. After the injection is completed the bird may be plucked and immersed in the preservative fluid in the same manner as in the preparation for dissection for muscles.

The arteries, the veins, and the nerves may now be dissected and studied in relation to one another and in relation to the muscles, the bones, and other structures. The skin should not be removed from the shanks till it is desired to dissect these parts, as the tissues quickly dry out. In fact, the tendons of the shanks and toes can best be dissected while the specimen is fresh.

A Study of the Structure of Bones.—Longitudinal and transverse sections of old bone may be made by making thin longitudinal and transverse sections with the bone saw, and then by making them very thin with a fine three-cornered file. Examining under the low power microscope, we note the lacunæ, the canaliculi, and the Haversian system.

Similar sections in green bone may be studied if prepared as follows. Secure specimen of bone just removed from a fowl and place it for three days in a 10 per cent. aqueous solution of hydrochloric acid. Test by puncturing with a needle, and, if all the mineral salts are removed, place in a water bath and wash for four hours. Pass it through the fluids usually employed in preparing specimens for sectioning with the microtome. Stain as sections of other tissue for microscopic study. See the description below. If the ends of the bone be included, it will enable the student to study not only compact bone but also cancellated bone and articular, or hyaline cartilage, and in some of the bones, as the femur, the red marrow.

Special Technic for the Dissection of Cranial and Spinal Nerves.—It is rather difficult to dissect the cranial and spinal nerves of the fowl, owing to the fact that the structures are very small. The bone is rather hard and the nerve tissue so delicate that great skill must be attained to achieve any degree of success.

A simple technic has been developed as follows: Place the head and neck, or other structures of the spinal column in a 10 per cent. aqueous solution of hydrochloric acid for three or more days, the time depending on the size of the specimen and the amount of soft structures surrounding it. This solution removes all the calcium salts from the bone and makes the removal of the bony structures a less difficult task.

Directions for the Study of Soft Structures.—Secure a specimen of the tissue to be studied—lung, muscle, intestine, liver, pancreas—from a normal fowl just killed. The specimen should be not more than ½ inch square. After first hardening three days in 10 per cent. formaldehyde. Pass through the following fluids:

1.	Alcohol, 95 per cent.	24 hours
2.	Alcohol, absolute	24 hours
3.	Alcohol and ether, equal parts	24 hours
4.	1 per cent. celloidin	24 hours
5.	2 per cent. celloidin	24 hours
6.	4 per cent. celloidin	24 hours
7.	6 per cent. celloidin	24 hours
8.	10 per cent. celloidin	24 hours
9.	Place on block, and as soon as solid, place in 80 per cent. alcohol until ready to section. Histoloid or parlodion will take the place of celloidin.	

In placing specimen on the block be careful that the specimen lies conveniently for cutting the sections in the right direction. As soon as the surface has hardened a little, add a few drops of thick celloidin, and repeat until there is a good body of celloidin. Allow to stand until the tissues are quite firmly fastened to the block, but not long enough to permit shrinking. Then place in 80 per cent. alcohol until the specimen is perfectly firm, 12 or more hours, before cutting.

All tissues, cut sections, and mounted blocks are to be placed in 80 per cent. alcohol. As containers for this purpose shell vials will be most handy.

Cut the sections with the microtome as thin as possible, the thinner the better. The following process of staining will make the nucleus blue and the cytoplasm reddish.

1. Float section in a tumbler of tap water.

2. Place section on slide, and immerse in hematoxylon for five to ten minutes.

3. Immerse in acid alcohol from two to five seconds.

5. Place on slide, and immerse in eosin from one-half to three minutes.

6. Wash thoroughly in alcohol.

7. Clarify in oil of cloves, oil of cedar, or beechwood creosote, ten minutes.

8. Mount in balsam.

9. Label and study.

Delafield's hematoxylon is prepared as follows:

Hematoxylon crystals	4 grams
Alcohol, 95 per cent	25 c.c.
Saturated aqueous solution of ammonia alum	400 c.c.

Add the hematoxylon dissolved in the alcohol to the alum solution, and expose in an unstoppered bottle to the light and air for three or four days. Filter and add:

| Glycerin | 100 c.c. |
| Alcohol, 95 per cent. | 100 c.c. |

Allow the solution to stand in the light until the color is sufficiently dark; then filter, and keep in a tightly stoppered bottle. The solution keeps well and is extremely powerful. So long as it is good the solution has a purplish tinge. If time permits, it would be wise to combine the alum, the hematoxylin, and the water, and to ripen the solution for two or three weeks before adding the other ingredients, which have a tendency to prevent oxidation.

The acid alcohol is made as follows:

| Absolute alcohol | 70 c.c. |
| Distilled water | 30 c.c. |

Mix.	
Above solution	99 c.c.
Hydrochloric acid	1 c.c.

Eosin is sold in two forms, that soluble in water and that soluble in alcohol. The eosin soluble in water is preferred, because with it a greater degree of diffusion in stain can be obtained.

Keep on hand a saturated aqueous solution and dilute with water as needed. The strength of the solution to be used varies somewhat with the tissue and the reagent in which it is to be fixed; but usually the strength should be between 1/10 and 1/2 per cent. when eosin is used after hematoxylin. The diluted solutions should contain 25 per cent. of alcohol, otherwise they will not keep well. When eosin is used before an aniline dye, such as methylene blue, a 5 per cent. or even a saturated solution should be used.

To Stain Sections of Liver for the Study of Kupffer Cells.—To bring out this reaction Keys suggests the following technic:

Fix small blocks of the fresh tissue of spleen or liver for eighteen to twenty-four hours in Müller's fluid plus 5 per cent. mercuric sublimate. Imbed in paraffin and section to 4 microns. Fix sections to slide, and stain twenty to forty minutes with acid carmine. Wash, and transfer to equal parts of a 2 per cent. aqueous solution of potassium ferrocyanid and of a 2 per cent. aqueous solution of hydrochloric acid. Remove after three to ten minutes, wash in distilled water, and pass quickly through a 0.5 per cent. aqueous erythrosin solution. Dehydrate in alcohol, clarify in xylol, and mount in Canada balsam.

To Prepare Anatomical Specimens for a Museum.—The Keiserling method gives the best results, since by this method the tissues retain their normal color. The three steps are as follows:

1. Place the specimen in the following solution and leave from one to seven days, the length of time depending upon the size of the specimen.

Formalin	200	c.c.
Potassium acetate	30	grams
Potassium nitrate	15	grams
Water	1000	c.c.

2. Pass the specimen through each of the following solutions, leaving it in each twenty-four hours or until the normal color is obtained. The specimen

should be removed from alcohol as soon as color is attained. If it is left in the alcohol too long it will again lose some of its color.

Alcohol	40 per cent.
Alcohol	60 per cent.
Alcohol	80 per cent.
Alcohol	full strength

3. Place the specimen in the following permanent solution, label, and place in museum.

Glycerin	40 c.c.
Potassium acetate	40 grams
Distilled water	400 c.c.

A small piece of thymol must be placed on the top of the liquid in each jar, or mold will develop and spoil the specimen.

To Make Specimens Transparent.—Specimens may be rendered transparent by the method of Spalteholz.

The steps are essentially as follows:

I. Preparation of the fresh tissue. If any parts are to be made conspicuous, as blood-vessels, or the lymphatic system, they must be injected with an insoluble, unbleachable substance. Spalteholz recommended carmine or methylene blue, and carbon. I have found Higgin's ordinary black carbon ink excellent for this purpose. The system or systems are injected with this substance while the tissue is perfectly fresh.

II. Fixation. The tissues are fixed preferably in 10 per cent. formalin. The length of time for fixation depends on the size of the tissue. In formalin it requires from eight hours upward to fix completely.

III. Rinsing. Running tap water for five or ten minutes accomplishes the rinsing. If it is impracticable to pass directly into the bleaching fluid, the tissues may be kept temporarily in 60 per cent. or 70 per cent. alcohol. They should then be rinsed again thoroughly in water when ready to bleach.

IV. Bleaching. The bleaching fluid used is hydrogen peroxid to which ammonia is added until a white precipitate forms. The proportion is approximately two parts of peroxid and one of ammonia. The material is bleached until all the color is removed and the tissue looks perfectly clear or slightly opaque. It may be quite transparent as the protein coagulate is somewhat white, but all yellow or reddish color should be bleached away.

V. Washing. Rinse thoroughly in running tap water or through several changes of water until the odor of ammonia is quite gone.

VI. Dehydration. Pass gradually through alcohol solutions, 30 per cent., 50 per cent., 60 per cent., 70 per cent., 80 per cent., 95 per cent., and last through absolute alcohol. The time required for dehydration will vary with the size of the material, but in either 95 per cent. solution or in absolute alcohol the tissue should remain until well hardened, twenty-four hours or more. Dehydration is completed by passing from absolute alcohol, to a fluid one-half absolute alcohol and one-half benzol; thence for a few days to pure benzol.

VII. Clarifying. From benzol the material is finally clarified in winter-green oil or any other standard clearer, and put up in its final position in the museum jar.

In commenting on this method Dr. A. F. Contant says: "I have found this to work very successfully in the injection of the lymphatics of the human skin, which, as you know, is a rather delicate injection and has been difficult of demonstration by other methods. With this method, however, I have prepared whole portions, as the side of the face, the leg, etc., of small animals and embryos which show clearly the whole course of the lymphatic vessels and nodes, in situ, and their relations to surrounding parts."

EQUIPMENT FOR THE DISSECTION LABORATORY

Figure 84 illustrates some essential equipment for the dissection laboratory. The equipment should consist of a pump, No. 1, which forces air through the rubber tube, No. 2, into the pressure tank, No. 4, which is guarded by the valve at No. 3. A dial, No. 5, indicates the pressure of the air within the tank. The air as needed is released through an outlet valve. The air now passes through the rubber tube, No. 6, into the left arm of the manometer at No. 7. The manometer is simply a glass tubing filled with mercury to the point indicated at No. 7. By the side of the opposite arm there is made a scale graduated in centimeters. The tube at No. 9 being attached to the Y-tube, conveys air under the same pressure into the injection chamber as that supporting the column of mercury of the manometer. It is necessary to have a pinch cock, as indicated at No. 11, to control the liquid within the tube. No. 12 illustrates the tube in the end of which, No. 13, there is inserted a trocar which is introduced into the artery and tightly tied with a cord.

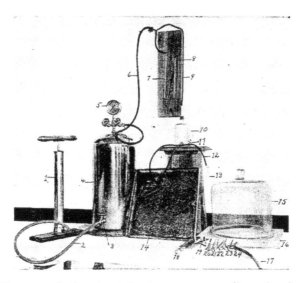

FIG. 84.—Photograph showing injection apparatus, dissection instruments, dissection pan and death chamber.

1, Pump. 2, Rubber tube leading from pump to compression chamber. 3, Stop cock. 4, Compression air tank. 5, Pressure indicator. 6, Rubber tube leading from pressure tank to manometer. 7, Top of mercury in manometer tube. 8, Open end of glass tube showing scale in centimeters. 9, Tube leading from 4 to injection chamber. 10, Injection chamber containing injection fluid. 11, Pinch cock on rubber tube connecting injection chamber with injection canula. 12, The tube. 13, Injection canula. 14, A metal dissection tray. 15, Death chamber. 16, Base of same. 17, Rubber tube connecting gas cock with death chamber. 18, Bone saw. 19, Bone cutter. 20, Scapula. 21, Tenaculum. 22, Forceps. 23, Straight scissors. 24, Small pair of curved scissors.

A convenient dissection tray may be made from galvanized iron. It should be 16 inches square and about 1 inch deep (Fig. 84, No. 14).

A death chamber is made by using a bell jar 12 inches in diameter at the bottom. This is used on a perfectly smooth board 16 inches square. Through the center of this board is inserted a glass tube ⅜ inch in diameter. To this glass tube is attached a rubber tube the other end of which is attached to a gas jet.

The instruments needed in dissecting are illustrated in Fig. 84. No. 18 is a bone saw; No. 20 is a scalpel; No. 21 is a tenaculum; No. 22 is a pair of forceps; No. 23 is a pair of straight scissors; and No. 24 is a small pair of curved scissors.

BIBLIOGRAPHY

BAUM AND ELLENBERGER, Handbuch der Vergleichenden Anatomie, Berlin, 1912.

BARROWS, H. R., *Me. E. S. Bull.* 232, pp. 16, plates 6, 1914.

BRADLEY, O. CHARNOCK, The Structure of the Fowl, pp. 146, illustrations 73.

CHAVEAU, A., Comparative Anatomy of Domestic Animals, New York, 1893.

CURTIS, M. R., Ligaments of the Oviduct of the Fowl, *Me. E. S. Bull.* 206, 1910.

DUVAL, ——, Atlas of Embryology.

FOSTER AND BALFOUR, Elements of Embryology.

GADOW, H., AND SELENKA, E., Vögel. I. Anatomischer Theil (Dr. H. G. Bronn's Klassen und Ordnungen des Thier-Reichs. Sechester Band, Vierte Ahtheiburg), Leipsig, 1891.

GRAY, HENRY, Anatomy Descriptive and Surgical, New York, 1908.

HEISLER, J. C., Embryology, Philadelphia, 1901.

KAUPP, B. F., Male Reproductive Organs of the Fowl, *Am. Jr. Vet. Med.*, vol. x, 11, 2.

KAUPP, B. F., Female Reproductive Organs of the Fowl, *Am. Vet. Rev.*, vol. 49, 11, 4.

Leisering Atlas of Anatomy, edited by W. ELLENBERGER, translated by A. T. PETERS, Chicago, 1905.

LILLIE, F. R., Embryology of the Chick, Chicago, 1906.

OWEN, R., Comparative Anatomy and Physiology of Vertebrates, London, 1866.

SCHMEISSER, H. C., A Study of the Blood of Fowls, Johns Hopkins Hospital, 1915.

SHUFELDT, R. W., Myology of the Raven, pp. 318, illustrations 76, London, 1890.

Strangeways' Veterinary Anatomy, edited by I. VAUGHAN, Edinburgh, 1892.

SURFACE, F. M., Histology of the Oviduct of the Fowl, *Me. E. S. Bull.*, 206, 1912.

www.ingramcontent.com/pod-product-compliance
Ingram Content Group UK Ltd.
Pitfield, Milton Keynes, MK11 3LW, UK
UKHW031827270325
456796UK00002B/256